The MPEG Representation of Digital Media

Leonardo Chiariglione
Editor

The MPEG Representation of Digital Media

Editor
Leonardo Chiariglione
CEDEO
Via Borgionera 103
10040 Villar Dora, Italy
leonardo@chiariglione.net

ISBN 978-1-4419-6183-9 e-ISBN 978-1-4419-6184-6
DOI 10.1007/978-1-4419-6184-6
Springer New York Dordrecht Heidelberg London

Library of Congress Control Number: 2011936224

Printed on acid-free paper

Springer is part of Springer Science+Business Media (www.springer.com)

Contents

Chapter 1
An Introduction to MPEG Digital Media

Leonardo Chiariglione

The Moving Picture Experts Group (MPEG) has produced a number of successful standards that have facilitated the conversion of the various components of the media industry from an analogue to the digital world. In the process MPEG has caused an orderly restructuring of the industry from a largely vertical arrangement to a horizontal one. This has been possible thanks to a properly designed standards development process that accommodates industry practices and promotes a healthy technology competition.

A communication system requires the means to convert a message to a form that is suitable for transmission over the selected channel. In a verbal communication acoustic waves are the channel and modulation of the waves in accordance to the intended verbal message is the means to convert the message to a form suitable for transmission over the channel. The channel used by sound recording of the early days was a medium (vinyl disc) that could be graved in proportion to the intensity of the sound and the means to convert the message was a membrane that converted sound intensity to the movement of a stylus. Analogue facsimile used a telephone line as the channel and a sensor detecting the intensity of the light from a page as the means to modulate the intensity of the telephone signal.

In the 170 years since the development of photography, the first media communication system that could communicate media information without human intervention, a very large number of such systems have been invented and deployed. As it is clear from the examples above, while the message can be considered as an abstract entity independent of the communication system, the means to convert the message to a form suitable for transmission over a selected channel is in general specific of that channel.

L. Chiariglione (✉)
CEDEO.net, Via Borgionera 103, 10040 Villar Dora, Turin, Italy
e-mail: leonardo@chiariglione.org

L. Chiariglione (ed.), *The MPEG Representation of Digital Media*,
DOI 10.1007/978-1-4419-6184-6_1, © Springer Science+Business Media, LLC 2012

The use of channels based on electrical, magnetic and electromagnetic techniques has made simpler the definition of a more generic form of "information representation" for communication. In the example above the microphone output can in principle be used for recording, transmission over a cable or over a radio channel. In general, however, the wildly differing specificities of the transmission mechanism have led designers of such communication systems to produce independent, "vertical" systems that are not able to communicate between them.

The development of digital technologies has triggered the birth of a new branch of science called Digital Signal Processing (DSP). Once electrical, magnetic, electromagnetic and even mechanical signals carrying information are converted into numbers, it is possible to convert the abstract entity called "message" into a digital form that is still "abstract" – because it does not depend on the channel – and can be processed by, stored in and communicated to physical devices for eventual conversion to a form that can be perceived by human senses.

Because of the designers' tendency to produce "vertical" solutions, analogue communication systems used to be beset with interoperability problems – suffice it to recall the incompatibilities between Beta and VHS video recorders or between the many colour television systems, the many versions of NTSC, PAL and SECAM. The existence of a digital representation of information could have led designers to produce more "horizontal" solutions, but the first digital communication systems designed, and some of them deployed, in the 1980s were still following the old pattern.

MPEG was established in 1988 in the framework of the International Organisation for Standardisation (ISO). The main driver for its creation was to exploit standardisation as the means to create markets for products and services with improved user experience because of assured interoperability. More than 20 years after that, one can indeed assess that MPEG standards for the digital representation of audio, video and related information have facilitated the transition of the media industry to the digital world, made convergence of hitherto disparate industries less chaotic than it could have been and sustains the growth of the industry at large.

Mindful of the need to create a layered organisation as opposed to the existing vertical organisation, MPEG has consistently applied its notion of "digital representation of information" to provide standard solutions that are, to the extent possible, independent of specific applications and industries. As a result its standards are indeed used by the broadest range of communication industries ever achieved in history and have been the drivers of the fastest ever transformation of an industry from a set of vertical systems to a set of horizontal layers.

This has been achieved by amalgamating industry practices into a set of coherent practices proper of the MPEG group. The typical process of MPEG standard development goes through the phases of gathering and rationalisation of requirements from multi-industry sources; the issuing of "Calls for proposals" of technologies or solutions satisfying some or all requirements; testing of and selection among suitable proposals; integration of proposed technologies into a complete solution; several rounds of ballots in the ISO National Bodies to perfect the solution; development of reference software, a complete implementation of the encoder (i.e. the part of the communication system that generates the digital representation of information) and of the decoder (i.e. the part that generates the information from its digital representation); development of conformance testing methodology to assess conformance of an implementation to the standard; verification testing to verify the degree of satisfaction of the standard to the original requirements; release of the agreed standard.

The process described requires that MPEG standards be anticipatory, in the sense that standards are developed anticipating needs that are not yet entirely defined. Obviously MPEG takes a risk but this is unavoidable if a standard technology is to be available before members of the industry make commitments to some other technology before a standard exists.

MPEG implements the ISO/IEC-defined standard development process in a very meticulous way, even adding some further steps. A typical MPEG standard goes through the phases of project identification, requirements definition, call for proposals, assessment of proposals, development of test model, working draft, committee draft, final committee draft, final draft international standard, verification test, conformance test. Therefore, when all steps are required and complex technologies are involved, it may take some time for MPEG to develop a standard.

Still the process is very healthy because it fosters competition among its members, but actually not just members since the rule in MPEG is that anybody is entitled to respond to a call for proposals (but membership is required to participate in the development of the standard). Letting companies compete with their technologies at the time a standard is being developed is very effective because competition happens at the time of important technology choices – those impacting on the performance of the standard – but also less wasteful because competition does not require the actual deployment of products or services that typically require investments orders of magnitude larger than those required by standardisation.

Since its early days, MPEG has discovered that targeting its standards to satisfy the needs of a broad range of industries is an obviously laudable but often nearly impossible task to achieve. Indeed requirements may be too diverse: an industry may seek maximum performance, another low cost and yet another flexibility. All this bearing in mind that convergence – an easy word for a complex phenomenon – may force different parts of different industries to coalesce.

MPEG has found a solution to this problem with the adoption of Profiles and Levels. With Profiles MPEG is able to control the amount of "technology" (i.e. features, performance and sometimes complexity) needed in an instance of the standard and with Levels MPEG is able to control the amount of "resources" (e.g. video resolution) involved.

Another feature of MPEG standards is the provision – whenever possible – of integrated solutions. MPEG prides itself for having been the first to develop a complete "system-level" standard solution for digital audio and video. The MPEG-1 standard includes a part for "video compression", another for "audio compression" and a third for "putting the two digital streams together" so that an application can deal with an "audio-visual" stream. This practice was followed with MPEG-2, MPEG-4 and MPEG-7 and is likely to continue for some of the planned standard.

MPEG was one of the first bodies to make intense use of programming languages in the actual design of its standards to the extent of writing significant parts of its standards in a pseudo C code. Therefore it was easy to make the further step of providing, next to its traditional textual form, computer code implementing a standard. In later standards the computer code has been given "normative" status, in the sense that "human readable" text and "machine readable" computer code are two alternative representations of the same standard.

Going against the practice of other bodies, MPEG has consistently applied the notion that standards should specify the minimum that is required for interoperability between implementations of transmitters (encoders) and receivers (decoders) from different sources. Therefore all MPEG standards contain clauses with normative value for decoders while the corresponding text for encoders has only informative value. The unique advantage of this policy is that there is a great deal of freedom in the design of encoders whose "performance" can improve over time partly as a result of the implementation of new ideas and partly from the continuous improvement of technology that enables more sophisticated implementations.

As MPEG standards are very sophisticated and offer a lot of freedom to the designer, sometimes incompatibilities between encoders and decoders from different sources arise. Continuing the established tradition of some industries (e.g. telecom, broadcasting and consumer electronics) all MPEG standards contain "conformance testing" clauses and suites to help implementors assess the conformity of their products to a standard before releasing them to the market.

Many MPEG standards carry the result of many years of research by companies participating in their development. Therefore MPEG is well aware that those who have invested in new successful technologies typically expect to be rewarded of their efforts that enable standards to provide leading-edge performance. MPEG develops its standards in accordance to the ISO/IEC/ITU Intellectual Property Rights (IPR) rules having as sole guidance the production of the most effective standards, with due regard to the state of technology.

As a body seeking to serve the needs of different industries across the board, MPEG sees it difficult and often improper to establish relationships with individual companies. On the other hand MPEG is fully aware that industries have felt and keep on feeling the need to establish organisations catering to their needs. MPEG has ongoing "liaison" relationships with some 50 organisations, within ISO, with the other two international standards organisations International Electrotechnical Commission (IEC) and International Telecommunication Union (ITU), and with a host of other bodies.

Of particular importance is the relationship with Study Group 16 (SG 16) of ITU. Since the time of the MPEG-2 Systems and Video standards (ISO/IEC 13818-1 and -2) MPEG has seen the benefit for the industry to exploit multi-industry standards jointly developed under the auspices of the international standardisation organisations – ISO, IEC and ITU – to serve the needs of their constituencies. This collaboration was resumed in 2001 with the goal of producing a new generation video coding standard – ISO/IEC 14496-10 Advanced Video Coding (AVC) and is currently progressing with the High-Efficiency Video Coding (HEVC) standard project.

Judging from the title of its mission “digital representation of media information”, the scope of MPEG may seem narrow. In practice, however, the scope is hardly so. Probably the area drawing most participants is “media compression”, which includes video, audio, 3D graphics (3DG) and other media types. A second area is “media composition”, namely the digital representation of “attributes” (time, space, interactivity, etc.) of the different “objects” in a multimedia scene. A third area is “media description”, namely the digital representation of the description – textual, but other descriptions as well – related to a media object. MPEG has several other activity areas that will not be described in this book. For sure the MPEG mission is highly strategic, judging from the some 500 participants in its quarterly meetings representing some 200 companies and organisations from some 25 countries.

After this introduction the book continues with three chapters dealing with past, present and future of MPEG video coding standards. Chapter 5 provides some samples of the huge world of MPEG “media description” standards focusing on video. The next three chapters deal with past, present and future of MPEG audio coding standards. Chapters 9 and 10 describe the task of “putting together” digital media for two types of MPEG standards. Chapter 11 introduces the world of standards for compression of 3DG information and is followed by Chap. 12 dealing with a new future-oriented approach to video and 3DG information coding. Chapter 13 deals with “subjective testing” of audio and video information that is used by MPEG throughout its entire process of video and audio coding standard development.

This book seeks to provide to its readers the means to achieve a basic technology understanding of the mechanisms underpinning the operation of MPEG standards for those making decisions in products and services based on digital media, those engaged in studies or developments of MPEG-related implementations starting from general background and those curious about such a wonderful developer of successful standard technologies as MPEG.

Chapter 2
MPEG Video Compression Basics

B.G. Haskell and A. Puri

2.1 Video Coding Basics

Video signals differ from image signals in several important characteristics. Of course the most important difference is that video signals have a camera frame rate of anywhere from 15 to 60 frames/s, which provides the illusion of smooth motion in the displayed signal.[1] Another difference between images and video is the ability to exploit temporal redundancy as well as spatial redundancy in designing compression methods for video. For example, we can take advantage of the fact that objects in video sequences tend to move in predictable patterns, and can therefore be *motion-compensated* from frame-to-frame if we can detect the object and its motion trajectory over time.

Historically, there have been five major initiatives in video coding [1–5] that have led to a range of video standards.

- Video coding for ISDN video teleconferencing, which has led to the ITU video coding standard called H.261 [6]. H.261 is also the baseline video mode for most multimedia conferencing systems.
- Video coding for low bitrate video telephony over POTS[2] networks with as little as 10 kbits/s allocated to video and as little as 5.3 kbits/s allocated to voice coding, which led to the ITU video coding standard called H.263 [7]. The H.263 low bitrate video codec is used at modem rates of from 14.4 to 56 kbits/s, where the modem rate includes video coding, speech coding, control information, and other logical channels for data.

---

[1] If the camera rate, chosen to portray motion, is below the display rate, chosen to avoid flicker, then some camera frames will have to be repeated.

[2] Plain Old Telephone Service.

B.G. Haskell (✉)
Apple Computer, 1 Infinite Loop, Cupertino, CA 95014, USA
e-mail: BGHaskell@comcast.net

L. Chiariglione (ed.), *The MPEG Representation of Digital Media*,
DOI 10.1007/978-1-4419-6184-6_2, © Springer Science+Business Media, LLC 2012

- Video coding for storing movies on CD-ROM with on the order of 1.2 Mbits/s allocated to video coding and 256 kbits/s allocated to audio coding, which led to the initial ISO MPEG-1 (Motion Picture Experts Group) standard [8].
- Video coding for broadband ISDN, broadcast and for storing video on DVD (Digital Video Disks) with on the order of 2–400 Mbits/s allocated to video and audio coding, which led to the ISO MPEG-2 video coding standard [9]. The ITU has given this standard the number H.262.
- Video coding for *object-based coding* at rates as low as 8 kbits/s, and as high as 1 Mbits/s, or higher, which led to the ISO MPEG-4 video coding standard [10]. Key aspects of this standard include independent coding of objects in a picture; the ability to interactively composite these objects into a scene at the display; the ability to combine graphics, animated objects, and natural objects in the scene; and finally the ability to transmit scenes in higher dimensionality formats (e.g., 3D).

Before delving in to details of standards, a few general remarks are in order. It is important to note that standards specify syntax and semantics of the compressed bit stream produced by the video encoder, and how this bit stream is to be parsed and decoded (i.e., decoding procedure) to produce a decompressed video signal. However, many algorithms and parameter choices in the *encoding* are not specified (such as motion estimation, selection of coding modes, allocation of bits to different parts of the picture, etc.) and are left open and depend greatly on encoder implementation. However it is a requirement that resulting bit stream from encoding be compliant to the specified syntax. The result is that the quality of standards based video codecs, even at a given bitrate, depends greatly on the encoder implementation. This explains why some implementations appear to yield better video quality than others.

In the following sections, we provide brief summaries of each of these video standards, with the goal of describing the basic coding algorithms as well as the features that support use of the video coding in multimedia applications.

2.1.1 Basics of Interframe Video Coding

A video scene captured as a sequence of frames can be efficiently coded by estimating and compensating for motion between frames prior to generating interframe difference signal for coding. Since motion compensation is a key element in most video coders, it is worthwhile understanding the basic concepts in this processing step.

For the ease of processing, each frame of video is uniformly partitioned into smaller units called Macroblocks (MBs, formally defined a bit later) where each macroblock consists of a 16×16 block of luma, and corresponding chroma blocks.

The way that the motion estimator works is illustrated in Fig. 2.1. Each block of pixels (say 16×16 luma block of a MB) in the current frame is compared with a set of candidate blocks of same size in the previous frame to determine the one that best predicts the current block. The set of blocks includes those within a search region in previous frame centered on the position of current block in the current frame.

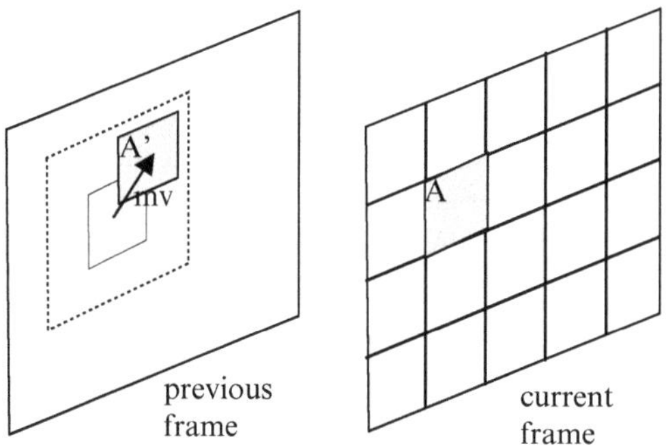

Fig. 2.1 Motion compensation of interframe blocks

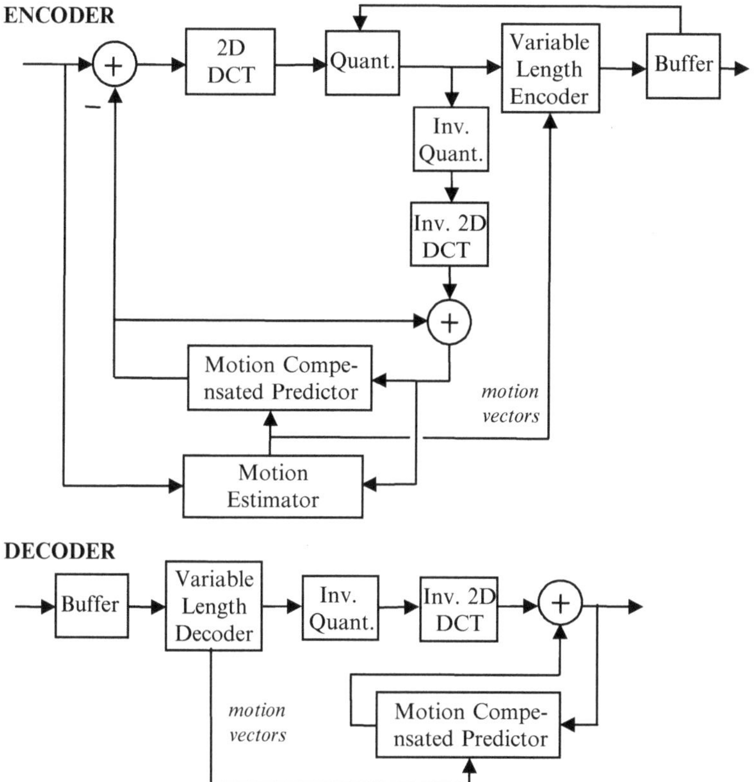

Fig. 2.2 Motion compensated encoder/decoder for interframe coding

When the best matching block is found, a motion vector is determined, which specifies the reference block.

Figure 2.2 shows a block diagram of a motion-compensated image codec. The key idea is to combine transform coding (in the form of the Discrete Cosine Transform (DCT) of 8×8 pixel blocks) with predictive coding (in the form of differential Pulse Code Modulation (PCM)) in order to reduce storage and computation of the compressed image, and at the same time to give a high degree of compression and adaptability.

Since motion compensation is difficult to perform in the transform domain, the first step in the interframe coder is to create a motion compensated prediction error in the pixel domain. For each block of current frame, a prediction block in the reference frame is found using motion vector found during motion estimation, and differenced to generate prediction error signal. This computation requires only a single frame store in the encoder and decoder. The resulting error signal is transformed using 2D DCT, quantized by an adaptive quantizer, entropy encoded using a Variable Length Coder (VLC) and buffered for transmission over a fixed rate channel.

We now discuss how various MPEG standards are built using principles and building blocks discussed so far.

2.2 The MPEG-1 Video Coding Standard

The MPEG-1 standard is the first true multimedia standard with specifications for coding, compression, and transmission of audio, video, and data streams in a series of synchronized, mixed Packets. The driving focus of the standard was storage of multimedia content on a standard CDROM, which supported data transfer rates of 1.4 Mb/s and a total storage capability of about 600 MB. MPEG-1 was intended to provide VHS VCR-like video and audio quality, along with VCR-like controls. MPEG-1 is formally called ISO/IEC 11172.

2.2.1 Requirements of the MPEG-1 Video Standard

Uncompressed digital video of full component TV resolution requires a very high transmission bandwidth, while VHS VCR-grade equivalent raw digital video requires transmission bandwidth of around 30 Mbits/s, with compression still necessary to reduce the bit-rate to suit most applications. The required degree of compression is achieved by exploiting the spatial and temporal redundancy present in a video signal. However, the compression process is inherently lossy, and the signal reconstructed from the compressed bit stream is not identical to the input video signal. Compression typically introduces some artifacts into the decoded signal.

The primary requirement of the MPEG-1 video standard was that it should achieve the high quality of the decoded motion video at a given bit-rate. In addition to picture quality under normal play conditions, different applications have additional requirements. For instance, multimedia applications may require the ability to randomly access and decode any single video picture[3] in the bitstream. Also, the ability to perform fast

[3] Frames and pictures are synonymous in MPEG-1.

search directly on the bit stream, both forward and backward, is extremely desirable if the storage medium has "seek" capabilities. It is also useful to be able to edit compressed bit streams directly while maintaining decodability. And finally, a variety of video formats were needed to be supported.

2.2.2 H.261 Coding Concepts as Applicable to MPEG-1 Video

The H.261 standard employs interframe video coding that was described earlier. H.261 codes video frames using a DCT on blocks of size 8×8 pixels, much the same as used for the original JPEG coder for still images. An initial frame (called an INTRA frame) is coded and transmitted as an independent frame. Subsequent frames, which are modeled as changing slowly due to small motions of objects in the scene, are coded efficiently in the INTER mode using a technique called Motion Compensation (MC) in which the displacement of groups of pixels from their position in the previous frame (as represented by so-called motion vectors) are transmitted together with the DCT coded difference between the predicted and original images.

2.2.2.1 H.261 Bitstream Data Hierarchy

We will first explain briefly the data structure in an H.261 video bit stream and then the functional elements in an H.261 decoder.

Only two picture formats, common intermediate format (CIF) and quarter-CIF (QCIF), are allowed. CIF pictures are made of three components: luminance Y and color differences Cb and Cr, as defined in ITU-R Recommendation BT601. The CIF picture size for Y is 352 pels[4] per line by 288 lines per frame. The two color difference signals are subsampled to 176 pels per line and 144 lines per frame. The image aspect ratio is 4(horizontal):3(vertical), and the picture rate is 29.97 non-interlaced frames per second. All H.261 standard codecs must be able to operate with QCIF; CIF is optional. A picture frame is partitioned into 8 line × 8 pel image blocks. A Macroblock (MB) is defined as four 8×8 (or one 16×16) Y block/s, one Cb block, and one Cr block at the same location.

The compressed H.261 video bit stream contains several layers. They are picture layer, group of blocks (GOB) layer, Macroblock (MB) layer, and block layer. The higher layer consists of its own header followed by a number of lower layers.

Picture Layer

In a compressed video bit stream, we start with the picture layer. Its header contains: Picture start code (PSC) a 20-bit pattern.

[4] Abbreviation of pixel.

Temporal reference (TR) a 5-bit input frame number.
Type information (PTYPE) such as CIF/QCIF selection.
Spare bits to be defined in later versions.

GOB Layer

At the GOB layer, a GOB header contains:

- Group of blocks start code (GBSC) a 16-bit pattern.
- Group number (GN) a 4-bit GOB address.
- Quantizer information (GQUANT) initial quantizer step size normalized to the range 1–31. At the start of a GOB, we set QUANT=GQUANT.
- Spare bits to be defined in later versions of the standard.

Next, comes the MB layer. An 11-bit stuffing pattern can be inserted repetitively right after a GOB header or after a transmitted Macroblock.

Macroblock (MB) Layer

At the MB layer, the header contains:

- Macroblock address (MBA) location of this MB relative to the previously coded MB inside the GOB. MBA equals one plus the number of skipped MBs preceding the current MB in the GOB.
- Type information (MTYPE) 10 types in total.
- Quantizer (MQUANT) normalized quantizer step size to be used until the next MQUANT or GQUANT. If MQUANT is received we set QUANT=MQUANT. Range is 1–31.
- Motion vector data (MVD) differential displacement vector.
- Coded block pattern (CBP) indicates which blocks in the MB are coded.

Blocks not coded are assumed to contain all zero coefficients.

Block Layer

The lowest layer is the block layer, consisting of quantized transform coefficients (TCOEFF), followed by the end of block (EOB) symbol. All coded blocks have the EOB symbol.

Not all header information need be present. For example, at the MB layer, if an MB is not Inter motion-compensated (as indicated by MTYPE), MVD does not exist. Also, MQUANT is optional. Most of the header information is coded using Variable Length Codewords.

There are essentially four types of coded MBs as indicated by MTYPE:

- Intra – original pels are transform-coded.
- Inter – frame difference pels (with zero-motion vectors) are coded. Skipped MBs are considered inter by default.

- Inter_MC – displaced (nonzero-motion vectors) frame differences are coded.
- Inter_MC_with_filter – the displaced blocks are filtered by a predefined loop filter, which may help reduce visible coding artifacts at very low bit rates.

2.2.2.2 H.261 Coding Semantics

A single-motion vector (horizontal and vertical displacement) is transmitted for one Inter_MC MB. That is, the four Y blocks, one Cb, and one Cr block all share the same motion vector. The range of motion vectors is +– 15 Y pels with integer values. For color blocks, the motion vector is obtained by halving the transmitted vector and truncating the magnitude to an integer value.

Motion vectors are differentially coded using, in most cases, the motion vector of the MB to the left as a prediction. Zero is used as a prediction for the leftmost MBs of the GOB, and also if the MB to the left has no motion vector.

The transform coefficients of either the original (Intra) or the differential (Inter) pels are ordered according to a zigzag scanning pattern. These transform coefficients are selected and quantized at the encoder, and then coded using variable-length codewords (VLCs) and/or fixed-length codewords (FLC), depending on the values. Just as with JPEG, successive zeros between two nonzero coefficients are counted and called a RUN. The value of a transmitted nonzero quantized coefficient is called a LEVEL. The most likely occurring combinations of (RUN, LEVEL) are encoded with a VLC, with the sign bit terminating the RUN-LEVEL VLC codeword.

The standard requires a compatible IDCT (inverse DCT) to be close to the ideal 64-bit floating point IDCT. H.261 specifies a measuring process for checking a valid IDCT. The error in pel values between the ideal IDCT and the IDCT under test must be less than certain allowable limits given in the standard, e.g., peak error <= 1, mean error <= 0.0015, and mean square error <= 0.02.

A few other items are also required by the standard. One of them is the image-block updating rate. To prevent mismatched IDCT error as well as channel error propagation, every MB should be intra-coded at least once in every 132 transmitted picture frames.

The contents of the transmitted video bit stream must also meet the requirements of the hypothetical reference decoder (HRD). For CIF pictures, every coded frame is limited to fewer than 256 Kbits; for QCIF, the limit is 64 Kbits, where K = 1,024. The HRD receiving buffer size is B + 256 Kbits, where $B = 4 \times R_{max}/29.97$ and R_{max} is the maximum connection (channel) rate. At every picture interval (1/29.97 s), the HRD buffer is examined. If at least one complete coded picture is in the buffer, then the earliest picture bits are removed from the buffer and decoded. The buffer occupancy, right after the above bits have been removed, must be less than B.

2.2.3 *MPEG-1 Video Coding*

Video coding as per MPEG-1 uses coding concepts similar to H.261 just described, namely spatial coding by taking the DCT of 8 × 8 pixel blocks, quantizing the DCT coefficients based on perceptual weighting criteria, storing the DCT coefficients for each block in a zigzag scan, and doing a variable run length coding of the resulting DCT coefficient stream. Temporal coding is achieved by using the ideas of uni- and bi-directional motion compensated prediction, with three types of pictures resulting, namely:

- *I* or Intra pictures which were coded independently of all previous or future pictures.
- *P* or Predictive pictures which were coded based on previous *I* or previous *P* pictures.
- *B* or Bi-directionally predictive pictures which were coded based on either the next and/or the previous pictures.

If video is coded at about 1.1 Mbits/s and stereo audio is coded at 128 kbits/s per channel, then the total audio/video digital signal will fit onto the CD-ROM bit-rate of approximately 1.4 Mbits/s as well as the North American ISDN Primary Rate (23 B-channels) of 1.47 Mbits/s. The specified bit-rate of 1.5 Mbits/s is not a hard upper limit. In fact, MPEG-1 allows rates as high as 100 Mbits/s. However, during the course of MPEG-1 algorithm development, coded image quality was optimized at a rate of 1.1 Mbits/s using progressive (*NonInterlaced*) scanned pictures.

Two *Source Input Formats* (SIF) were used for optimization. One corresponding to NTSC was 352 pels, 240 lines, 29.97 frames/s. The other corresponding to PAL, was 352 pels, 288 lines, 25 frames/s. SIF uses 2:1 color subsampling, both horizontally and vertically, in the same 4:2:0 format as H.261.

2.2.3.1 Basics of MPEG-1 Video Compression

Both spatial and temporal redundancy reduction are needed for the high compression requirements of MPEG-1. Most techniques used by MPEG-1 have been described earlier.

Exploiting Spatial Redundancy

The compression approach of MPEG-1 video combines elements of JPEG, elements of H.261, and significant new elements that allow not only higher compression but also frequent entry points into the video stream.

Because video is a sequence of still images, it is possible to achieve some compression using techniques similar to JPEG. Such methods of compression are called intraframe coding techniques, where each picture of video is individually and independently compressed or encoded. Intraframe coding exploits the spatial redundancy that exists between adjacent pels of a picture. Pictures coded using only intraframe coding are called *I-pictures*.

As in JPEG and H.261, the MPEG-1 video-coding algorithm employs a block-based two-dimensional DCT. A picture is first divided into 8×8 blocks of pels, and the two-dimensional DCT is then applied independently on each block. This operation results in an 8×8 block of DCT coefficients in which most of the energy in the original (pel) block is typically concentrated in a few low-frequency coefficients. The coefficients are scanned and transmitted in the same zigzag order as JPEG and H.261.

A quantizer is applied to the DCT coefficients, which sets many of them to zero. This quantization is responsible for the lossy nature of the compression algorithms in JPEG, H.261 and MPEG-1 video. Compression is achieved by transmitting only the coefficients that survive the quantization operation and by entropy-coding their locations and amplitudes.

Exploiting Temporal Redundancy

Many of the interactive requirements can be satisfied by intraframe coding. However, as in H.261, the quality achieved by intraframe coding alone is not sufficient for typical video signals at bit-rates around 1.1 Mbits/s.

Temporal redundancy results from a high degree of correlation between adjacent pictures. The MPEG-1 algorithm exploits this redundancy by computing an interframe difference signal called the prediction error. In computing the prediction error, the technique of motion compensation is employed to correct for motion. A Macroblock (MB) approach is adopted for motion compensation.

In unidirectional or *Forward Prediction*, 16×16 luma block of each macroblock in the current picture to be coded is matched with a block of the same size in a previous picture called the Reference picture. As in H.261 blocks of the Reference picture that "best match" the 16×16 luma blocks of current picture, are called the *Prediction* blocks. The prediction error is then computed as the difference between the Target block and the Prediction block.[5] The position of this best-matching Prediction block is indicated by a motion vector that describes the displacement between it and the Target block. Unlike H.261 where each motion vector is specified at "integer pel" accuracy, in MPEG-1 each motion vector is specified at "half-pel" accuracy, thus allowing improved prediction. The motion vector information is also encoded and transmitted along with the prediction error. Pictures coded using Forward Prediction are called *P-pictures*.

The prediction error itself is transmitted using the DCT-based intraframe encoding technique summarized above. In MPEG-1 video (as in H.261), motion compensation is performed on MBs (16×16 luma and associated chroma), representing a reasonable trade-off between the compression provided by motion compensation and the cost associated with transmitting the motion vectors.

Bidirectional Temporal Prediction

Bidirectional temporal prediction, also called Motion-Compensated Interpolation, is a key feature of MPEG-1 video. Pictures coded with Bidirectional prediction use

[5] Prediction 16×16 blocks do not, in general, align with coded 16×16 luma (of MB) boundaries in the Reference frame.

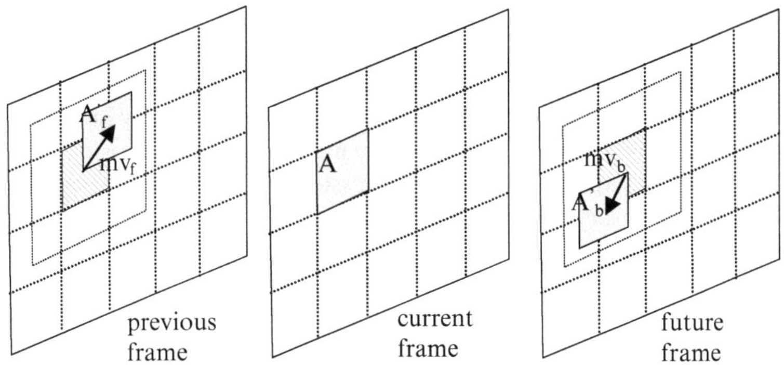

Fig. 2.3 Motion compensated interpolation in a bidirectionally predicted picture

two Reference pictures, one in the past and one in the future. A Target 16×16 luma block in bidirectionally coded pictures can be predicted by a 16×16 block from the past Reference picture (Forward Prediction), or one from the future Reference picture (Backward Prediction), or by an average of two 16×16 luma blocks, one from each Reference picture (Interpolation). In every case, a Prediction 16×16 block from a Reference picture is associated with a motion vector, so that up to two motion vectors per macroblock may be used with Bidirectional prediction. As in the case of unidirectional prediction, motion vectors are represented at "half-pel" accuracy. Motion-Compensated Interpolation for a 16×16 block in a Bidirectionally predicted "current" frame is illustrated in Fig. 2.3.

Pictures coded using Bidirectional Prediction are called *B-pictures*. Pictures that are Bidirectionally predicted are never themselves used as Reference pictures, i.e., Reference pictures for B-pictures must be either P-pictures or I-pictures. Similarly, Reference pictures for P-pictures must also be either P-pictures or I-pictures.

Bidirectional prediction provides a number of advantages. The primary one is that the compression obtained is typically higher than can be obtained from Forward (unidirectional) prediction alone. To obtain the same picture quality, Bidirectionally predicted pictures can be encoded with fewer bits than pictures using only Forward prediction.

However, Bidirectional prediction does introduce extra delay in the encoding process, because pictures must be encoded out of sequence. Further, it entails extra encoding complexity because block matching (the most computationally intensive encoding procedure) has to be performed twice for each Target block, once with the past Reference picture and once with the future Reference picture.

2.2.3.2 MPEG-1 Bitstream Data Hierarchy

The MPEG-1 video standard specifies the syntax and semantics of the compressed bit stream produced by the video encoder. The standard also specifies how this bit stream is to be parsed and decoded to produce a decompressed video signal.

The details of the motion estimation matching procedure are not part of the standard. However, as with H.261 there is a strong limitation on the variation in bits/picture in the case of constant bit-rate operation. This is enforced through a Video Buffer Verifier (VBV), which corresponds to the Hypothetical Reference Decoder of H.261. Any MPEG-1 bit stream is prohibited from overflowing or underflowing the buffer of this VBV. Thus, unlike H.261, there is no picture skipping allowed in MPEG-1.

The bit-stream syntax is flexible in order to support the variety of applications envisaged for the MPEG-1 video standard. To this end, the overall syntax is constructed in a hierarchy[6] of several Headers, each performing a different logical function.

Video Sequence Header

The outermost Header is called the Video Sequence Header, which contains basic parameters such as the size of the video pictures, Pel Aspect Ratio (PAR), picture rate, bit-rate, assumed VBV buffer size and certain other global parameters. This Header also allows for the optional transmission of JPEG style Quantizer Matrices, one for Intra coded pictures and one for Non-Intra coded pictures. Unlike JPEG, if one or both quantizer matrices are not sent, default values are defined. Private user data can also be sent in the Sequence Header as long as it does not contain a Start Code Header, which MPEG-1 defines as a string of 23 or more zeros.

Group of Pictures (GOP) Header

Below the Video Sequence Header is the Group of Pictures (GOP) Header, which provides support for random access, fast search, and editing. A sequence of transmitted video pictures is divided into a series of GOPs, where each GOP contains an intra-coded picture (I-picture) followed by an arrangement of Forward predictive-coded pictures (P-pictures) and Bidirectionally predicted pictures (B-pictures).

Figure 2.4 shows a GOP example with six pictures, 1–6. This GOP contains I-picture 1, P-pictures 4 and 6, and B-pictures 2, 3 and 5. The encoding/transmission order of the pictures in this GOP is shown at the bottom of Fig. 2.4. B-pictures 2 and 3 are encoded after P-picture 4, using P-picture 4 and I-picture 1 as reference. Note that B-picture 7 in Fig. 2.4 is part of the next GOP because it is encoded after I-picture 8.

Random access and fast search are enabled by the availability of the I-pictures, which can be decoded independently and serve as starting points for further decoding. The MPEG-1 video standard allows GOPs to be of arbitrary structure and length.

---

[6] As in H.261, MPEG-1 uses the term Layers for this hierarchy. However, Layer has another meaning in MPEG-2. Thus, to avoid confusion we will not use Layers in this section.

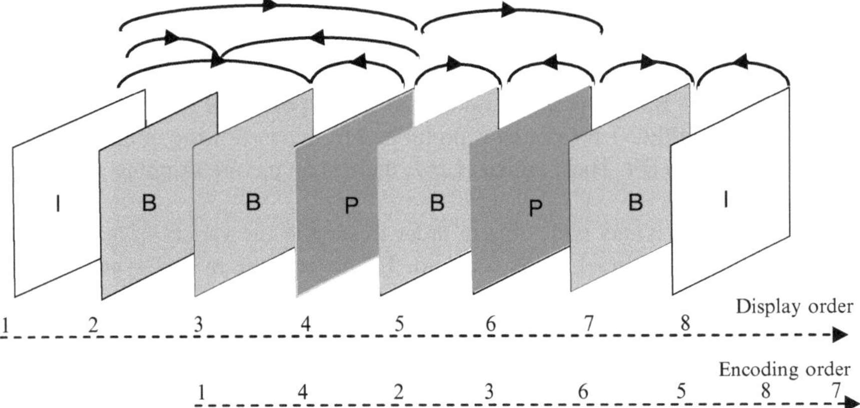

Fig. 2.4 Illustration of motion compensated coding of frames

Picture Header

Below the GOP is the Picture Header, which contains the type of picture that is present, e.g., I, P or B, as well as a Temporal Reference indicating the position of the picture in display order within the GOP. It also contains a parameter called vbv_delay that indicates how long to wait after a random access before starting to decode. Without this information, a decoder buffer could underflow or overflow following a random access.

Slice Header

A *Slice* is a string of consecutive Macroblocks of arbitrary length running from left to right and top to bottom across the picture. The Slice Header is intended to be used for re-synchronization in the event of transmission bit errors. Prediction registers used in the differential encoding of motion vectors and DC Intra coefficients are reset at the start of a Slice. It is again the responsibility of the encoder to choose the length of each Slice depending on the expected bit error conditions. The first and last MBs of a Slice cannot be skipped MBs, and gaps are not allowed between Slices. The Slice Header contains the vertical position of the Slice within the picture, as well as a quantizer_scale parameter (corresponding to GQUANT in H.261).

Macroblock Header

The Macroblock (MB) is the 16×16 motion compensation unit. In the Macroblock Header, the horizontal position (in MBs) of the first MB of each Slice is coded with the MB Address VLC. The positions of additional transmitted MBs are coded

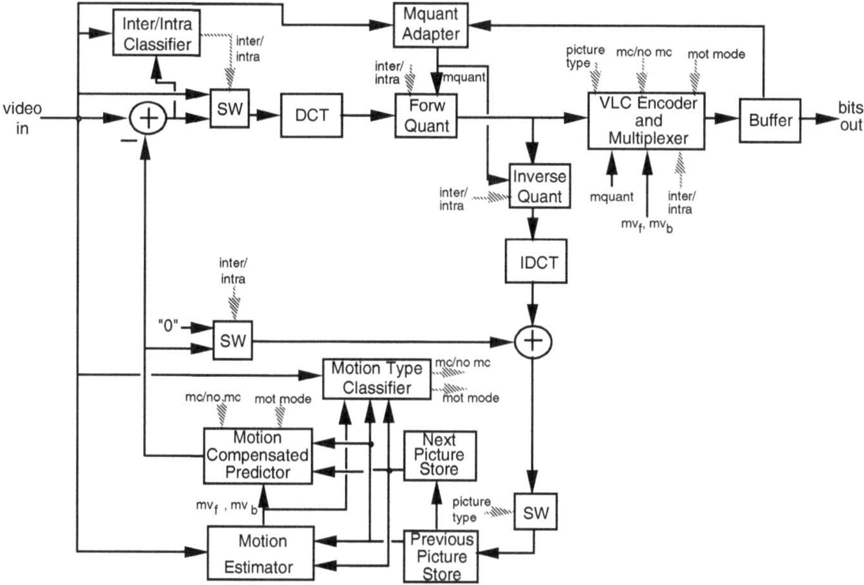

Fig. 2.5 MPEG-1 video encoder

differentially with respect to the most recently transmitted MB, also using the MB Address VLC. Skipped MBs are not allowed in I-pictures. In P-pictures, skipped MBs are assumed NonIntra with zero coefficients and zero motion vectors. In B-pictures, skipped MBs are assumed NonIntra with zero coefficients and motion vectors the same as the previous MB. Also included in the Macroblock Header are MB Type (Intra, NonIntra, etc.), quantizer_scale (corresponding to MQUANT in H.261), motion vectors and coded block pattern. As with other Headers, these parameters may or may not be present, depending on MB Type.

Block

A *Block* consists of the data for the quantized DCT coefficients of an 8×8 Block in the Macroblock. It is VLC coded as described in the next sections. For noncoded Blocks, the DCT coefficients are assumed to be zero.

2.2.3.3 MPEG-1 Video Encoding

Figure 2.5 shows a typical MPEG-1 video encoder. It is assumed that frame reordering takes place before coding, i.e., I- or P-pictures used for B-picture prediction must be coded and transmitted before any of the corresponding B-pictures.

Input video is fed to a Motion Compensation Estimator/Predictor that feeds a prediction to the minus input of the Subtractor. For each MB, the Inter/Intra Classifier then compares the input pels with the prediction error output of the Subtractor. Typically, if the mean square prediction error exceeds the mean square pel value, an Intra MB is decided. More complicated comparisons involving DCT of both the pels and the prediction error yield somewhat better performance, but are not usually deemed worth the cost.

For Intra MBs the prediction is set to zero. Otherwise, it comes from the Predictor, as described above. The prediction error is then passed through the DCT and Quantizer before being coded, multiplexed and sent to the Buffer.

Quantized Levels are converted to reconstructed DCT coefficients by the Inverse Quantizer and then inverse transformed by the IDCT to produce a coded prediction error. The Adder adds the prediction to the prediction error and clips the result to the range 0–255 to produce coded pel values.

For B-pictures the Motion Compensation Estimator/Predictor uses both the Previous Picture and the Future Picture. These are kept in picture stores and remain unchanged during B-picture coding. Thus, in fact, the Inverse Quantizer, IDCT and Adder may be disabled during B-picture coding.

For I and P-pictures the coded pels output by the Adder are written to the Future Picture Store, while at the same time the old pels are copied from the Future Picture Store to the Previous Picture Store. In practice this is usually accomplished by a simple change of memory addresses.

The Coding Statistics Processor in conjunction with the Quantizer Adapter control the output bit-rate in order to conform to the Video Buffer Verifier (VBV) and to optimize the picture quality as much as possible. A simple control that works reasonably well is to define a target buffer fullness for each picture in the GOP. For each picture the quantizer_scale value is then adjusted periodically to try to make the actual buffer fullness meet the assigned value. More complicated controls would, in addition, exploit spatio-temporal masking in choosing the quantizer_scale parameter for each MB.

2.2.3.4 MPEG-1 Video Decoding

Figure 2.6 shows a typical MPEG-1 video decoder. It is basically identical to the pel reconstruction portion of the encoder. It is assumed that frame reordering takes place after decoding and video output. However, extra memory for reordering can often be avoided if during a write of the Previous Picture Store, the pels are routed also to the display.

The decoder cannot tell the size of the GOP from the bitstream parameters. Indeed it does not know until the Picture Header whether the picture is I-, P- or B-type. This could present problems in synchronizing audio and video were it not for the Systems part of MPEG-1, which provides Time Stamps for audio, video and ancillary data. By presenting decoded information at the proper time as indicated by the Time Stamps, synchronization is assured.

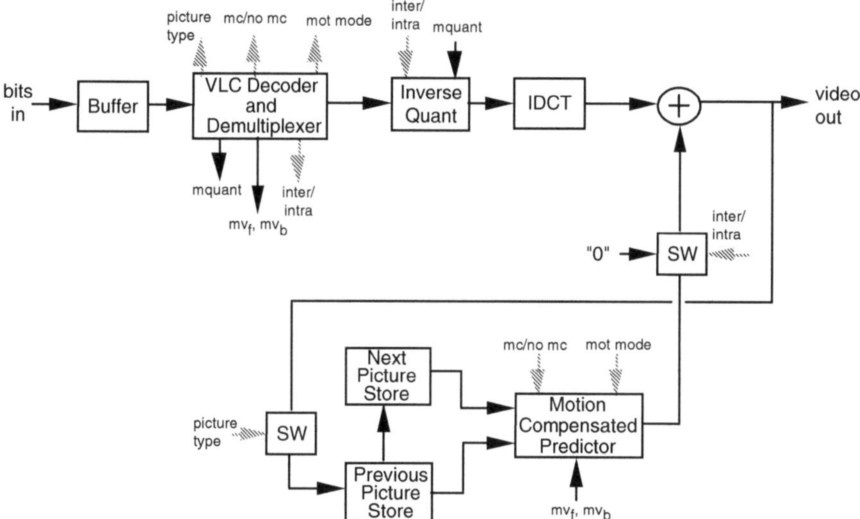

Fig. 2.6 MPEG-1 video decoder

Decoders often must accommodate occasional transmission bit errors. Typically, when errors are detected either through external means or internally through the arrival of illegal data, the decoder replaces the data with skipped MBs until the next Slice is detected. The visibility of errors is then rather low unless they occur in an I-picture, in which case they may be very visible throughout the GOB. A cure for this problem was developed for MPEG-2 and will be described later.

2.2.4 MPEG-1 Capabilities and Interoperability

In demonstrations of MPEG-1 video at a bit-rate of 1.1 Mbits/s, SIF resolution pictures have been used. This resolution is roughly equivalent to one field of an interlaced NTSC or PAL picture. The quality achieved by the MPEG-1 video encoder at this bit-rate has often been compared to that of VHS[7] videotape playback.

Although the MPEG-1 video standard was originally intended for operation in the neighborhood of the above bit-rate, a much wider range of resolution and bit-rates is supported by the syntax. The MPEG-1 video standard thus provides a generic bit-stream syntax that can be used for a variety of applications. MPEG-1 video (Part 2 of ISO/IEC 11172) provides all the details of the syntax, complete with informative sections on encoder procedures that are outside the scope of the standard.

[7] VHS, an abbreviation for Video Home System, is a registered trademark of the Victor Company of Japan.

To promote interoperability between MPEG-1 applications (bitstreams) and decoders, MPEG introduced the concept of constrained parameters; the following parameters were specified.

Horizontal size <= 768 pels, vertical size of <=576 lines, Picture rate <=30
Number of MBs/pic <= 396, number of MBs/s <=396 × 25, Bitrate <=4,640

2.3 The MPEG-2 Video Coding Standard

The MPEG-2 standard was designed to provide the capability for compressing, coding, and transmitting high quality, multi-channel, multimedia signals over broadband networks, for example using ATM (Asynchronous Transmission Mode) protocols. The MPEG-2 standard specifies the requirements for video coding, audio coding, systems coding for combining coded audio and video with user-defined private data streams, conformance testing to verify that bitstreams and decoders meet the requirements, and software simulation for encoding and decoding of both the program and the transport streams. Because MPEG-2 was designed as a transmission standard, it supports a variety of Packet formats (including long and variable length Packets from 1 kB up to 64 kB), and provides error correction capability that is suitable for transmission over cable TV and satellite links.

MPEG-2 video was aimed at video bit-rates above 2 Mbits/s. Specifically, it was originally designed for high quality encoding of interlaced video from standard TV with bitrates on the order of 4–9 Mb/s. As it evolved, however, MPEG-2 video was expanded to include high resolution video, such as High Definition TV (HDTV). MPEG-2 video was also extended to include hierarchical or scalable coding. The official name of MPEG-2 is ISO/IEC 13818.

The original objective of MPEG-2 was to code interlaced BT601 video at a bit-rate that would serve a large number of consumer applications, and in fact one of the main differences between MPEG-1 and MPEG-2 is that MPEG-2 handles interlace efficiently. Since the picture resolution of BT601 is about four times that of the SIF of MPEG-1, the bit-rate chosen for MPEG-2 optimization were 4, and 9 Mbits/s. However, MPEG-2 allows much higher bitrates.

A bit-rate of 4 Mbits/s was deemed too low to enable high quality transmission of every BT601 color sample. Thus, an MPEG-2 4:2:0 format was defined to allow for 2:1 vertical subsampling of the color, in addition to the normal 2:1 horizontal color subsampling of BT601. Pel positions are only slightly different than the CIF of H.261 and the SIF of MPEG-1.

For interlace, the temporal integrity of the chrominance samples must be maintained. Thus, MPEG-2 normally defines the first, third, etc. rows of 4:2:0 chrominance CbCr samples to be temporally the same as the first, third, etc. rows of luminance Y samples. The second, fourth, etc. rows of chrominance CbCr samples are temporally the same as the second, fourth, etc. rows of luminance Y samples. However, an override capability is available to indicate that the 4:2:0 chrominance samples are all temporally the same as the temporally first field of the frame.

At higher bit-rates the full 4:2:2 color format of BT601 may be used, in which the first luminance and chrominance samples of each line are co-sited. MPEG-2 also allows for a 4:4:4 color format.

2.3.1 Requirements of the MPEG-2 Video Standard

The primary requirement of the MPEG-2 video standard was that it should achieve high compression of interlaced video while maintaining high video quality. In addition to picture quality, different applications have additional requirements even beyond those provided by MPEG-1. For instance, multipoint network communications may require the ability to communicate simultaneously with SIF and BT601 decoders. Communication over packet networks may require prioritization so that the network can drop low priority packets in case of congestion. Broadcasters may wish to send the same program to BT601 decoders as well as to progressive scanned HDTV decoders. In order to satisfy all these requirements MPEG-2 has defined a large number of capabilities.

However, not all applications require all the features of MPEG-2. Thus, to promote interoperability amongst applications, MPEG-2 introduced several sets of algorithmic choices (*Profiles*) and choice of constrained parameters (*Levels*) to enhance interoperability.

2.3.2 MPEG-2 Video Coding

As with MPEG-1, both spatial and temporal redundancy reduction are needed for the high compression requirements of MPEG-2. For progressive scanned video there is very little difference between MPEG-1 and MPEG-2 compression capabilities. However, interlace presents complications in removing both types of redundancy, and many features have been added to deal specifically with it. MP@ML[8] only allows 4:2:0 color sampling.

For interlace, MPEG-2 specifies a choice of two picture structures. *Field Pictures* consist of fields that are coded independently. With *Frame Pictures*, on the other hand, field pairs are combined into frames before coding. MPEG-2 requires interlace to be decoded as alternate top and bottom fields.[8] However, either field can be temporally first within a frame.

MPEG-2 allows for progressive coded pictures, but interlaced display. In fact, coding of 24 frame/s film source with 3:2 pulldown display for 525/60 video is also supported.

[8] The top field contains the top line of the frame. The bottom field contains the second (and bottom) line of the frame.

2.3.2.1 Basics of MPEG-2 Video Compression

We now provide a brief overview of basic principles employed in MPEG-2 video compression.

Exploiting Spatial Redundancy

As in MPEG-1, the MPEG-2 video-coding algorithm employs an 8×8 Block-based two-dimensional DCT. A quantizer is applied to each DCT coefficient, which sets many of them to zero. Compression is then achieved by transmitting only the non-zero coefficients and by entropy-coding their locations and amplitudes.

The main effect of interlace in frame pictures is that since alternate scan lines come from different fields, vertical correlation is reduced when there is motion in the scene. MPEG-2 provides two features for dealing with this.

First, with reduced vertical correlation, the zigzag scanning order for DCT coefficients used for progressive frames is no longer optimum. Thus, MPEG-2 has an *Alternate Scan* that may be specified by the encoder on a picture by picture basis. A threshold on the sum of absolute vertical line differences is often sufficient for deciding when to use the alternate scan. Alternatively, DCT coefficient amplitudes after transformation and quantization could be examined.

Second, a capability for field coding within a frame picture MB is provided. That is, just prior to performing the DCT, the encoder may reorder the luminance lines within a MB so that the first 8 lines come from the top field, and the last 8 lines come from the bottom field. This reordering is undone after the IDCT in the encoder and the decoder. Chrominance blocks are not reordered in MP@ML. The effect of this reordering is to increase the vertical correlation within the luminance blocks and thus decrease the DCT coefficient energy. Again, a threshold on the sum of absolute vertical line differences is often sufficient for deciding when to use field DCT coding in a MB.

Exploiting Temporal Redundancy

As in MPEG-1, the quality achieved by intraframe coding alone is not sufficient for typical MPEG-2 video signals at bit-rates around 4 Mbits/s. Thus, MPEG-2 uses all the temporal redundancy reduction techniques of MPEG-1, plus other methods to deal with interlace.

Frame Prediction for Frame Pictures

MPEG-1 exploits temporal redundancy by means of motion compensated MBs and the use of P-pictures and B-pictures. MPEG-2 refers to these methods as *Frame Prediction for Frame Pictures*, and, as in MPEG-1, assigns up to one motion vector to Forward Predicted Target MBs and up to two motion vectors to Bidirectionally Predicted Target MBs. Prediction MBs are taken from previously coded Reference frames, which must be either P-pictures or I-pictures. Frame Prediction works well for slow to moderate motion, as well as panning over a detailed background. Frame Prediction cannot be used in Field Pictures.

Field Prediction for Field Pictures

A second mode of prediction provided by MPEG-2 for interlaced video is called *Field Prediction for Field Pictures*. This mode is conceptually similar to Frame Prediction, except that Target MB pels all come from the same field. Prediction MB pels also come from one field, which may or may not be of the same polarity (top or bottom) as the Target MB field. For P-pictures, the Prediction MBs may come from either of the two most recently coded I- or P-fields. For B-pictures, the Prediction MBs are taken from the two most recently coded I- or P-frames. Up to one motion vector is assigned to Forward Predicted Target MBs and up to two motion vectors to Bidirectionally Predicted Target MBs. This mode is not used for Frame Pictures and is useful mainly for its simplicity.

Field Prediction for Frame Pictures

A third mode of prediction for interlaced video is called Field Prediction for Frame Pictures [9]. With this mode, the Target MB is first split into top-field pels and bottom-field pels. Field prediction, as defined above, is then carried out independently on each of the two parts of the Target MB. Thus, up to two motion vectors are assigned to Forward Predicted Target MBs and up to four motion vectors to Bidirectionally Predicted Target MBs. This mode is not used for Field Pictures and is useful mainly for rapid motion.

Dual Prime for P-Pictures

A fourth mode of prediction, called *Dual Prime for P-pictures*, transmits one motion vector per MB. From this motion vector two preliminary predictions are computed, which are then averaged together to form the final prediction. The previous picture cannot be a B-picture for this mode. The first preliminary prediction is identical to Field Prediction, except that the Prediction pels all come from the previously coded fields having the same polarity (top or bottom) as the Target pels. Prediction pels, which are obtained using the transmitted MB motion vector, come from one field for Field Pictures and from two fields for Frame Pictures.

2.3.2.2 MPEG-2 Bitstream Data Hierarchy

Most of MPEG-2 consists of additions to MPEG-1. The Header structure of MPEG-1 was maintained with many additions within each Header. For constant bit-rates, the Video Buffer Verifier (VBV) was also kept. However, unlike MPEG-1, picture skipping as in H.261 is allowed.

Video Sequence Header

The Video Sequence Header contains basic parameters such as the size of the coded video pictures, size of the displayed video pictures, Pel Aspect Ratio (PAR), picture rate, bit-rate, VBV buffer size, Intra and NonIntra Quantizer Matrices (defaults are the same as MPEG-1), Profile/Level Identification, Interlace/Progressive Display Indication, Private user data, and certain other global parameters.

Group of Pictures (GOP) Header

Below the Video Sequence Header is the Group of Pictures (GOP) Header, which provides support for random access, fast search, and editing. The GOP Header contains a time code (hours, minutes, seconds, frames) used by certain recording devices. It also contains editing flags to indicate whether the B-pictures following the first I-picture can be decoded following a random access.

Picture Header

Below the GOP is the Picture Header, which contains the type of picture that is present, e.g., I, P or B, as well as a *Temporal Reference* indicating the position of the picture in display order within the GOP. It also contains the parameter vbv_delay that indicates how long to wait after a random access before starting to decode. Without this information, a decoder buffer could underflow or overflow following a random access.

Within the Picture Header a picture display extension allows for the position of a *display rectangle* to be moved on a picture by picture basis. This feature would be useful, for example, when coded pictures having Image Aspect Ratio (IAR) 16:9 are to be also received by conventional TV having IAR 4:3. This capability is also known as *Pan and Scan.*

MPEG-2 Slice Header

The *Slice* Header is intended to be used for re-synchronization in the event of transmission bit errors. It is the responsibility of the encoder to choose the length of each Slice depending on the expected bit error conditions. Prediction registers used in differential encoding are reset at the start of a Slice. Each horizontal row of MBs must start with a new Slice, and the first and last MBs of a Slice cannot be skipped. In MP@ML, gaps are not allowed between Slices. The Slice Header contains the vertical position of the Slice within the picture, as well as a quantizer_scale parameter (corresponding to GQUANT in H.261). The Slice Header may also contain an indicator for Slices that contain only Intra MBs. These may be used in certain Fast Forward and Fast Reverse display applications.

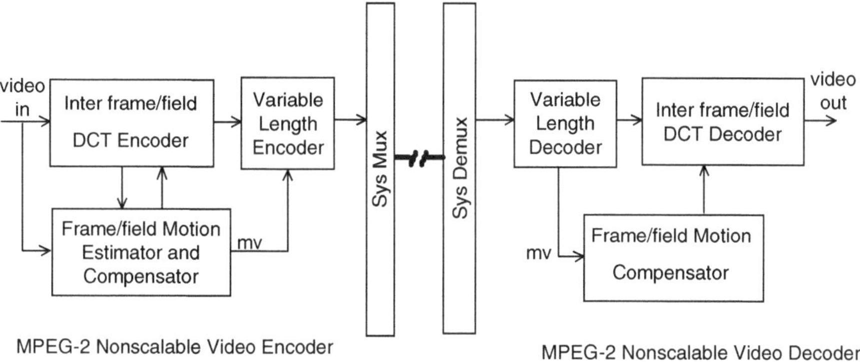

Fig. 2.7 A generalized codec for MPEG-2 nonscalable video coding

Macroblock Header

In the *Macroblock* Header, the horizontal position (in MBs) of the first MB of each Slice is coded with the MB Address VLC. The positions of additional transmitted MBs are coded differentially with respect to the most recently transmitted MB also using the MB Address VLC. Skipped MBs are not allowed in I-pictures. In P-pictures, skipped MBs are assumed NonIntra with zero coefficients and zero motion vectors. In B-pictures, skipped MBs are assumed NonIntra with zero coefficients and motion vectors the same as the previous MB. Also included in the Macroblock Header are MB Type (Intra, NonIntra, etc.), Motion Vector Type, DCT Type, quantizer_scale (corresponding to MQUANT in H.261), motion vectors and coded block pattern. As with other Headers, many parameters may or may not be present, depending on MB Type.

MPEG-2 has many more MB Types than MPEG-1, due to the additional features provided as well as to the complexities of coding interlaced video. Some of these will be discussed below.

Block

A *Block* consists of the data for the quantized DCT coefficients of an 8×8 Block in the Macroblock. It is VLC coded as described. MP@ML has six blocks per MB. For noncoded Blocks, the DCT coefficients are assumed to be zero.

2.3.2.3 MPEG-2 Encoding/Decoding Architecture

Figure 2.7 shows a high-level block diagram of an MPEG-2 nonscalable video codec. The video encoder consists of an inter-frame/field DCT encoder, a frame/field motion estimator and compensator, and a variable length entropy encoder (VLE). The frame/field DCT encoder exploits spatial redundancies in the video,

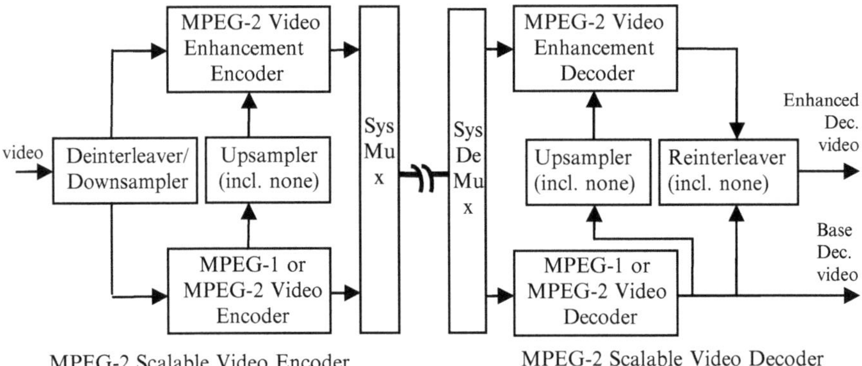

Fig. 2.8 A generalized coder for MPEG-2 scalable video coding

and the frame/field motion compensator exploits temporal redundancies in the video signal. The coded video bitstream is sent to a systems multiplexer, Sys Mux, which outputs either a Transport or a Program Stream.

The MPEG-2 decoder of Fig. 2.7 consists of a variable length entropy decoder (VLD), inter-frame/field DCT decoder, and the frame/field motion compensator. Sys Demux performs the complementary function of Sys Mux and presents the video bitstream to the VLD for decoding of motion vectors and DCT coefficients. The frame/field motion compensator uses a motion vector decoded by the VLD to generate a motion compensated prediction that is added back to a decoded prediction error signal to generate decoded video out. This type of coding produces nonscalable video bitstreams, since normally the full spatial and temporal resolution coded is the one that is expected to be decoded.

2.3.2.4 MPEG-2 Scalable Encoding/Decoding Architecture

A high level block diagram of a generalized codec for MPEG-2 scalable video coding is shown in Fig. 2.8. Scalability is the property that allows decoders of various complexities to be able to decode video of resolution/quality commensurate with their complexity from the same bit-stream. The generalized structure of Fig. 2.8 provides capability for both spatial and temporal resolution scalability in the following manner. The input video goes through a pre-processor that produces two video signals, one of which (called the Base Layer) is input to a standard MPEG-1 or MPEG-2 nonscalable video encoder, and the other (called the Enhancement Layer) is input to an MPEG-2 enhancement video encoder. The two bitstreams, one from each encoder, are multiplexed in Sys Mux (along with coded audio and user data). In this manner it becomes possible for two types of decoders to be able to decode a video signal of quality commensurate with their complexity, from the same encoded bit stream.

Table 2.1 MPEG-2 video profile/level subset of interest

	High	1,920 pels/frame 1,152 lines/frame 60 frames/s 62.7 Msamples/s 80 Mbits/s
Level	High-1,440	1,440 pels/frame 1,152 lines/frame 60 frames/s 47.0 Msamples/s 60 Mbits/s
	Main	720 pels/frame 576 lines/frame 30 frames/s 10.4 Msamples/s 15 Mbits/s
	Main profile	

2.3.3 MPEG-2 Capabilities and Interoperability

A *Profile* specifies the coding features supported, while a *Level*, specifies the picture resolutions, bit-rates, etc. that can be handled. A number of Profile-Level combinations have been defined, the most important ones are *Main Profile at Main Level* (MP@ML), and Main Profile at High Levels (MP@HL-1,440, and MP@HL). Table 2.1 shows parameter constraints for MP@ML, MP@HL-1,440, and MP@HL.

In addition to the Main Profile, several other MPEG-2 Profiles have been defined. The *Simple* Profile is basically the Main Profile without B-pictures. The *SNR Scalable* Profile adds SNR Scalability to the main Profile. The *Spatially Scalable* Profile adds Spatial Scalability to the SNR Scalable Profile, and the *High* Profile adds 4:2:2 color to the Spatially Scalable Profile. All scalable Profiles are limited to at most three layers.

The Main Level is defined basically for BT601 video. In addition, the *Simple* Level is defined for SIF video. Two higher levels for HDTV are the *High-1,440* Level, with a maximum of 1,440 pels/line, and the *High* Level, with a maximum of 1,920 pels/line.

2.4 The MPEG-4 (Part 2) Video Standard

MPEG-4 builds on and combines elements from three fields: digital television, interactive graphics and the World Wide Web with focus of not only providing higher compression but also an increased level of interactivity with the audio-visual content, i.e., access to and manipulation of objects in a visual scene.

The MPEG-4 standard, or formally ISO 14496 was started in 1993, and core of the standard was completed by 1998. Since then there have been a number of extensions and additions to the standard.

Overall, MPEG-4 addresses the following main requirements:

- Content Based Access
- Universal Accessibility
- Higher Compression

While previous standards code pre-composed media (e.g., video and corresponding audio), MPEG-4 codes individual visual objects and audio objects in the scene and delivers, in addition, a coded description of the scene. At the decoding side, the scene description and individual media objects are decoded, synchronized and composed for presentation.

2.4.1 H.263 Coding Concepts as Applicable to MPEG-4 Video

The H.263 video coding standard is based on the same DCT and motion compensation techniques as used in H.261. Several small, incremental improvements in video coding were added to the H.263 baseline standard for use in POTS conferencing. These included the following:

- Half pixel motion compensation in order to reduce the roughness in measuring best matching blocks with coarse time quantization. This feature significantly improves the prediction capability of the motion compensation algorithm in cases where there is object motion that needs fine spatial resolution for accurate modeling.
- Improved variable-length coding.
- Reduced overhead of channel input.
- Optional modes including unrestricted motion vectors that are allowed to point outside the picture.
- Arithmetic coding in place of the variable length (Huffman) coding.
- Advanced motion prediction mode including overlapped block motion compensation.
- A mode that combines a bidirectionally predicted picture with a normal forward predicted picture.

In addition, H.263 supports a wider range of picture formats including 4CIF (704×576 pixels) and 16CIF ($1,408 \times 1,152$ pixels) to provide a high resolution mode picture capability.

2.4.2 MPEG-4 Video Objects Coding

While the MPEG-4 Visual coding includes coding of natural video as well as synthetic visual (graphics, animation) objects, here we discuss coding of natural video objects only.

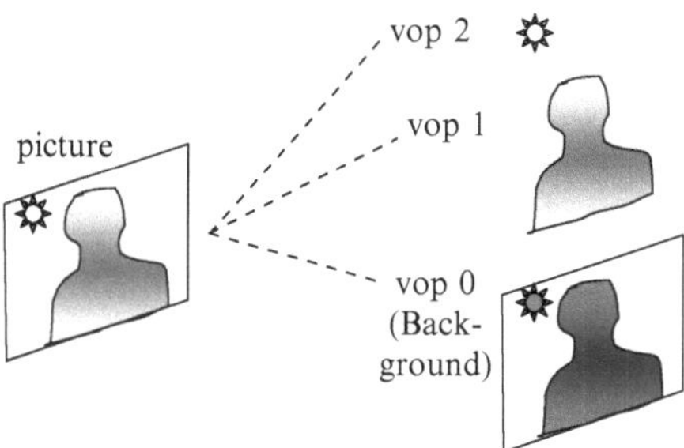

Fig. 2.9 Segmentation of a picture into spatial vops

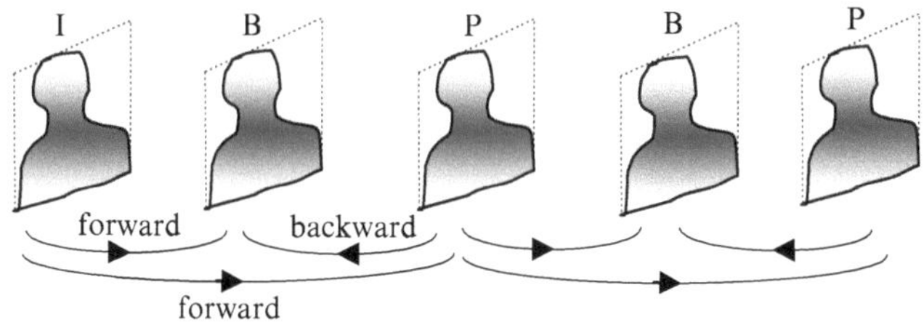

Fig. 2.10 An example prediction structure when using I, P and B-vops

The concept of Video Objects (VO) and their temporal instances, Video Object Planes (VOPs) is central to MPEG-4 video. A temporal instance of a video object can be thought of as a snapshot of arbitrary shaped object that occurs within a picture, such that like a picture, it is intended to be an access unit, and, unlike a picture, it is expected to have a semantic meaning.

Figure 2.9 shows an example of segmentation of a Picture into spatial VOPs with each VOP representing a snapshot in time of a particular object.

2.4.2.1 Basics of MPEG-4 (Part 2) Compression

To consider how coding takes place as per MPEG-4 video, consider a sequence of vops. MPEG-4 extends the concept of intra (I-) pictures, predictive (P-) and bidirectionally predictive (B-) pictures of MPEG-1/2 video to vops, thus I-vop, P-vop and B-vop result. Figure 2.10 shows a coding structure that uses a B-vop between a pair of reference vops (I- or P-vops). In MPEG-4, a vop can be represented by its shape (boundary), texture (luma and chroma variations), and motion (parameters).

The extraction of shape of vops per picture of an existing video scene by semi-automatic or automatic segmentation, results in a *binary shape* mask. Alternatively, if special mechanisms such as blue-screen composition are employed for generation of video scenes, either binary, or *grey scale* (8-bit) shape can be extracted. Both the binary- or the grey scale- shape mask needs to be sufficiently compressed for efficient representation. The shape information is encoded separate from encoding of the texture field, and motion parameters.

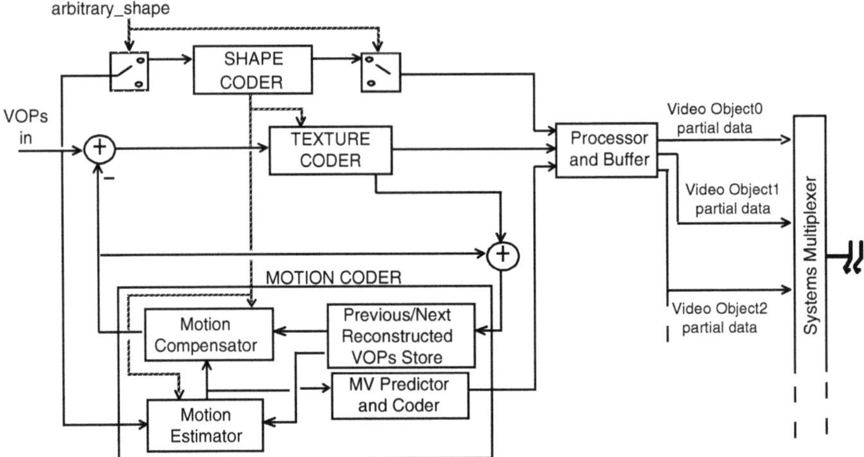

Fig. 2.11 MPEG-4 video encoder

2.4.2.2 MPEG-4 (Part 2) Video Compression Core Tools, and Architecture

In Fig. 2.11, we show the high level structure of an MPEG-4 Video Encoder which codes a number of video objects of a scene. Its main components are: Motion Coder, Texture and Shape Coder. The Motion Coder performs motion estimation and compensation of macroblock and blocks, similar to H.263 and MPEG-1/2 but modified to work with arbitrary shapes. The Texture Coder uses block DCT coding that is much more optimized than that of H.263 and MPEG-1/2, and further it is adapted to work with arbitrary shapes. The Shape Coder, represents an entirely new component not present in previous standards. The coded data of multiple vops (representing multiple simultaneous video objects) is buffered and sent to the System Multiplexer.

The Shape Coder allows efficient compression of both the binary-, and the grey-scale shape. Coding of binary shape is performed on a 16×16 block basis as follows. To encode a shape, a tight bounding rectangle that extends to multiples of 16×16 blocks is first created around the shape, with transparency of extended samples set to zero. On a macroblock basis, binary shape is then encoded using context information, motion compensation, and arithmetic coding. Similarly, grey scale shape is also encoded on a macroblock basis. Grey scale shape has similar properties as luma, and is thus encoded in a similar manner using texture and motion compensation tools used for coding of luma.

The Motion Coder consists of a Motion Estimator, Motion Compensator, Previous/Next Vops Store and Motion Vector (MV) Predictor and Coder. In case of P-vops, Motion Estimator computes motion vectors using the current vop and temporally previous reconstructed vop available from the Previous Reconstructed Vops Store. In case of B-vops, Motion Estimator computes motion vectors using the current vop and temporally previous reconstructed vop from the Previous Reconstructed Vop Store, as well as, the current vop and temporally next vop from the Next Reconstructed Vop Store. The Motion Compensator uses these motion vectors to compute motion compensated prediction signal using the temporally previous reconstructed version of the same vop (reference vop). The MV Predictor and Coder generates prediction for the MV to be coded. Special padding is needed for motion compensation of arbitrary shaped Vops.

MPEG-4 B-pictures include an efficient mode for generating predictions, called the "temporal direct mode" or simply the "direct mode". This mode allows motion vectors not only for 16×16 blocks but also for 8×8 blocks. The mode uses direct bi-directional motion compensation by computing scaled forward and backward motion vectors. The direct mode utilizes the motion vectors of the co-located macroblock in the most recently decoded I- or P-vop; the co-located macroblock is defined as the macroblock that has the same horizontal and vertical index with the current macroblock in the B-vop. If the co-located macroblock is transparent and thus the motion vectors are not available, these motion vectors are assumed to be zero. Besides the direct mode, MPEG-4 B-vops also support other modes of MPEG-1/2 B-pictures such as forward, backward, and bidirectional (interpolation) modes.

The Texture Coder codes luma and chroma blocks of macroblocks of a vop. Two types of macroblocks exist, those that lie inside the vop and those that lie on the boundary of the vop. The blocks that lie inside the vop are coded using DCT coding similar to that used in H.263 but optimized in MPEG-4. The blocks that lie on the vop boundary are first padded and then coded similar to the block that lie inside the vop. The remaining blocks that are inside the bounding box but outside of the coded vop shape are considered transparent and thus are not coded.

The texture coder uses block DCT coding and codes blocks of size 8×8 similar to H.263 and MPEG-1/2, with the difference that since vop shapes can be arbitrary, the blocks on the vop boundary require padding prior to texture coding. The general operations in the texture encoder are: DCT on original or prediction error blocks of size 8×8, quantization of 8×8 block DCT coefficients, scanning of quantized coefficients and variable length coding of quantized coefficients. For inter (prediction error block) coding, the texture coding details are similar to that of H.263 and MPEG-1/2. However, for intra coding a number of techniques to improve coding efficiency such as variable de-quantization, prediction of the DC and AC coefficients, and adaptive scanning of coefficients, are included.

2.4.2.3 MPEG-4 (Part 2) Video Compression Extension Tools

A number of advanced coding efficiency tools were added to the core standard and can be listed as follows.

Quarter pel prediction allows higher precision in interpolation during motion compensation resulting in higher overall coding efficiency. While it involves higher coding cost of motion information it also allows for more accurate representation of motion, significantly reducing motion compensated prediction errors.

Global motion compensation allows increase in coding efficiency due to compact parametric representation of motion that can be used to efficiently compensate for global movement of object/s or that of camera. It supports the five transformation models for the warping - stationary, translational, isotropic, affine, and perspective. The pixels in a global motion compensated macroblock are predicted using global motion prediction. The predicted macroblock is obtained by applying warping to the reference object. Each macroblock can be predicted either from previous vop by global motion compensation using warping parameters, or by local motion compensation using macroblock/block motion vectors. The resulting prediction error block is encoded by DCT coding.

Sprite in MPEG-4 video, is a video object that represents a region of coherent motion. Thus a sprite is a single large composite image resulting from blending of uncovered pixels belonging to various temporal instances of the video object. Alternatively, given a sprite, video objects at various temporal instances can be synthesized from it. A sprite is thus said to have the potential of capturing spatiotemporal information in a compact manner. MPEG-4 supports coding of sprites as a special type of vop (S-vop).

Shape-Adaptive DCT (SA-DCT) was added to the MPEG-4 standard to provide a more efficient way of coding blocks that contain shape boundaries and thus need to be padded resulting in much higher number of coefficients as compared to the active pixels of such blocks that need to be coded. SA-DCT produces at the most the same number of DCT coefficients as the number of pixels of a block that need to be coded. It is applied as a normal separable DCT, albeit on rows and columns such that each row or columns is not of equal size. Coefficients of SA-DCT are then quantized and coded as other DCT coefficients.

2.4.2.4 MPEG-4 (Part 2) Video Error Resilience Tools

MPEG-4 includes a number of error resilience tools to increase robustness of coded data. First, MPEG-4 uses a packet approach to allow resynchronization of the bitstream to restart decoding after detection of errors. It involves introducing periodic resynchronization markers in the bitstream. The length of the packet is independent of the contents of a packet but depends only on a fixed number of bits, ie, resynchronization markers are distributed uniformly throughout the bitstream. Second it allows for insertion of header extension information in the bitstream that makes it possible to decode each video packet independent of the previous information, allowing for faster resynchronization. Third, it allows for increased error resilience via use of "data partitioning" that involves separation of motion and macroblock header information from texture information using a motion marker to identify the separation point. This allows for better error concealment as in case of residual errors in the texture information, the previously received motion information can be used to provide motion compensated error concealment. Fourth, it allows use of reversible variable length codes (RVLCs) for DCT coefficient coding that can allow faster recovery after errors have been detected as the bitstream can be decoded both in forward and reverse directions. Thus the part of the bitstream that can not be decoded in the forward direction can be decoded in the backward direction; this

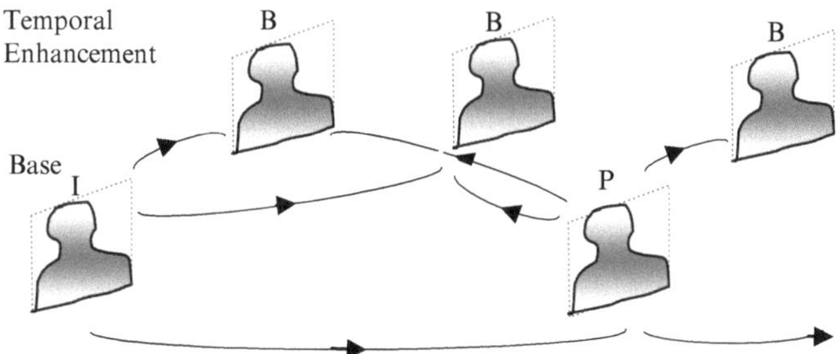

Fig. 2.12 An example of temporally scalable coding of vops

allows for recovery of a larger portion of the bitstream and thus increases the correctly recovered data in case of errors. However the use of RVLC incurs some additional constant penalty due to over-head so it can decrease coding performance when no errors occur.

2.4.2.5 MPEG-4 (Part 2) Video Scalability Tools

Some applications may require video objects to be simultaneously available in several spatial or temporal resolutions. Thus, MPEG-4 supports Temporal, and Spatial scalability of video objects.

Temporal Scalability

Temporal scalability involves partitioning of the video objects in layers, where the base layer provides the basic temporal rate, and the enhancement layers, when combined with the base layer, provide higher temporal rates. MPEG-4 supports temporal scalability of both rectangular and arbitrary shaped video objects. Temporal scalability is both computationally simple and efficient (essentially similar to B-vops coding that is already supported in nonscalable video coding).

Figure 2.12 shows an example of two layer temporally scalable coding of arbitrary shaped vops. In this example, I- and P-vops in the base layer are independently coded using prediction from previous decoded vops, while the enhancement layer consists of B-vops that are coded using prediction from decoded base layer vops.

Spatial Scalability

Spatial scalability involves generating two spatial resolution video layers from a single source such that the lower layer can provide the lower spatial resolution, and the higher spatial resolution can be obtained by combining the enhancement layer with the interpolated lower resolution base layer. MPEG-4 supports spatial scalability of both rectangular and arbitrary shaped objects. Since arbitrarily shaped objects can have varying sizes and locations, the video objects are resampled. An absolute

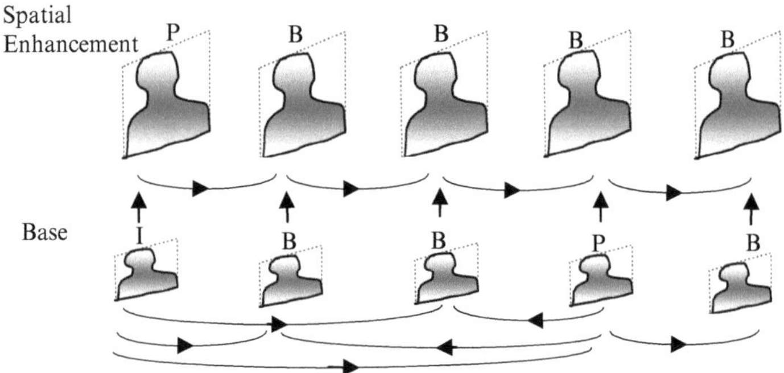

Fig. 2.13 An example of spatially scalable coding of vops

reference coordinate frame is used to form the reference video objects. The resampling involves temporal interpolation, and spatial relocalization. This ensures the correct formation of the spatial prediction used to compute the spatial scalability layers. Before the object are resampled, the objects are padded to form objects that can be used in the spatial prediction process.

Figure 2.13 shows an example of two layer spatially scalable coding of arbitrary shaped vops. In this example, the I-, P- and B-vops in the base layer are independently coded using prediction from previous decoded vops in the same layer while the enhancement layer P- and B-vops are coded using prediction from decoded base layer vops as well as prediction from previous decoded vops of enhancement layer.

2.4.3 MPEG-4 Video Capabilities and Interoperability

As mentioned earlier, MPEG-4 is a large standard containing many tools as it was aimed at addressing a new class of applications requiring access to coded audio-visual objects. In this chapter we have covered key aspects of MPEG-4 part2 video including video compression core tools, video compression extension tools, error resilience tools, and scalability tools. Due to large number of tools, there are a large number of (Object Type) Profiles in MPEG-4 each with many levels. However only the Simple Profile, and the Advanced Simple Profile have been widely implemented.

Besides MPEG-4 part 2 video, later MPEG-4 added a newer more efficient video coding technology referred to as MPEG-4 part 10 video (also known as AVC/H.264 standard) and is discussed in the next chapter of this book.

2.5 Summary

In this chapter we have provided an overview of the MPEG standards including requirements, technology, and applications/profiles of the MPEG-1 video, MPEG-2 video and MPEG-4 Part 2 video standards. Table 2.2 presents a side-side comparison of the technology features of the three early MPEG standards.

Table 2.2 Comparison of MPEG-1, MPEG-2, and MPEG-4 part 2 video standards

Features/technology	MPEG-1 video	MPEG-2 video	MPEG-4 (part 2) video
1. Video structure			
Sequence type	Progressive	Interlace, progressive	Interlace, progressive
Picture structure	Frame	Top/bot. field, frame	Top/bot. field, frame
2. Core compression			
Picture types	I-, P-, B-pictures	I-, P-, B-pictures	I-, P-, B-rect./arb. vops
Motion comp. block size	16×16	16×16 (16×8 field)	16×16, 8×8 (overlapped)
B-motion comp modes	Forw/backw/intp	Forw/backw/intp	Forw/backw/intp/direct
Transform, and block size	DCT, 8×8	DCT, 8×8	DCT, 8×8
Quantizer type	Linear	Linear and nonlinear	Linear and nonlinear
Quantizer matrices	Full	Full	Full and partial
Coefficient prediction	DC only	DC only	Adapt. DC and AC coeff.
Scan of coefficients	Zigzag	Zigzag, alt. vert	Zigzag, alt. vert, alt. horiz
Entropy coding	Fixed vlc	Intra, and inter vlc	Adaptive vlc
3. Higher compression			
Impr. motion comp. prec.	No	No	Quarter pel (v2)
Impr. motion comp. mode	No	No	Global motion comp. (v2)
Background representation	No	No	Sprite mode (v2)
Transform of nonrect. block	No	No	Shape adaptive DCT (v2)
4. Interlace compression			
Frame and field pictures	Frame picture	Frame, field picture	Frame, field picture
Motion comp. frame/field	No	No	Yes
Coding frame/field	No	No	Yes
5. Object based video			
Arbitrary object shape	No	No	Vop
Shape coding	No	No	Binary, grey scale alpha
6. Error resilient video			
Resynch markers	Slice start code	Slice start code	Flexible markers
Data partitioning	No	Coeff only	Coeffs, mv
Resilient entropy coding	–	–	Reversible vlc
7. Scalable video			
SNR scalability	No	Picture	(Via spatial)
Spatial scalability	No	Picture	Picture, vop
Temporal scalability	No	Picture	Picture, vop
8. Stereo/3D/multiview			
Stereo/multiview support	No	Yes	Yes

References

1. D. J. LeGall, "MPEG: A Video Compression Standard for Multimedia Applications," Communications of the ACM, Vol. 34, No.4, April 1991, pp. 47–58.
2. "Digital Video," IEEE Spectrum, Vol. 29, No.3, March 1992.
3. J. L. Mitchell, W. B. Pennebaker, C. E. Fogg, and D. J. LeGall, *MPEG Video Compression Standard*, Chapman and Hall, New York, 1997.
4. B. G. Haskell, A. Puri, and A. N. Netravali, *Digital Video: An Introduction to MPEG-2*, Chapman and Hall, New York, 1997.
5. A. Puri, and T. Chen, *Multimedia Systems, Standards and Networks*, Marcel Dekker, New York, 2000.
6. ITU-T Recommendation H.261, Video Codec for Audiovisual Services at p*64 kbits/sec.
7. ITU-T Recommendation H.263, Video Coding for Low Bit rate Communication.
8. "Coding of Moving Pictures and Associated Audio for Digital Storage Media up to about 1.5Mbit/s," ISO/IEC 11172-2: Video (November 1991).
9. "Generic Coding of Moving Pictures and Associated Audio Information," ISO/IEC 13818-2: Video (November 1994).
10. "Coding of Audio-Visual Objects," ISO/IEC 14496-2: Video (October 1998).
11. A. Puri, R. Aravind, and B. G. Haskell, "Adaptive Frame/Field Motion Compensated Video Coding", Signal Processing Image Communication, Vol. 1–5, pp. 39–58, Feb. 1993.

Chapter 3
MPEG Video Compression Advances

Jens-Rainer Ohm and Gary J. Sullivan

3.1 Overview and Background

Today's flagship in the MPEG fleet of video coding standards is the Advanced Video Coding (AVC) standard [1]. This chapter is devoted to a discussion of the AVC design and its capabilities. The AVC standard has achieved a significant improvement in compression capability compared to prior standards, and it provides a network-friendly representation of video that addresses both non-conversational (storage, broadcast, or streaming) and conversational (videotelephony) applications. Extensions of AVC have given it efficient support for additional functionality such as scalability at the bitstream level and 3D stereo/multiview coding.

The AVC standard was jointly developed by MPEG (the ISO/IEC Moving Picture Experts Group) and the ITU-T Video Coding Experts Group (VCEG). It is published both as ISO/IEC International Standard 14496-10 (informally known as MPEG-4 Part 10) and ITU-T Recommendation H.264.

Certainly, at this point more than one billion devices worldwide have been built which are running AVC. Some of its prominent applications include mobile videotelephony; TV and video players; HDTV broadcast and storage; and internet video (download and conferencing).

The main steps of AVC technical development were done in five phases:

- The first version finalized in 2003
- *Fidelity range extensions* (FRExt) finalized in 2004
- *Professional profiles* (PP) extensions finalized in 2007
- *Scalable video coding* (SVC) extensions finalized in 2007
- *Multi-view video coding* (MVC) extensions finalized in 2008/2009

G.J. Sullivan (✉)
Microsoft Corp, Redmond, WA, USA
e-mail: garysull@microsoft.com

L. Chiariglione (ed.), *The MPEG Representation of Digital Media*,
DOI 10.1007/978-1-4419-6184-6_3, © Springer Science+Business Media, LLC 2012

Whereas the first three developments mainly targeted compression performance for consumer and professional uses, SVC and MVC have provided additional functionality such as variable-rate access at the bitstream level and support for multiple-camera captures in combination with efficient compression.

3.2 Concepts

As has been the case for all ITU-T and ISO/IEC video coding standards since H.261 in 1990, only the central decoder has been standardized, by imposing restrictions on the bitstream and syntax, and defining the decoding process of the syntax elements such that every decoder conforming to the standard will produce similar or identical output when given an encoded bitstream that conforms to the constraints of the standard. The AVC standard actually specifies the decoder output exactly, so that all decoders will produce the same results. This provides an assurance of decoded picture quality and eases conformance testing. Additional post-processing of the decoded output pictures is allowed and is considered outside the scope of the standard, as are techniques for handling losses or corruption of the compressed data.

To address the need for flexibility and customizability, the AVC design covers a *video coding layer* (VCL), which is designed to efficiently represent the video content, and a *network abstraction layer* (NAL), which formats the VCL representation of the video and provides header information in a way that enables the coded video to be conveyed by a variety of transport layers or storage media (see Fig. 3.1). The most important top-level control data is carried in the *parameter sets* – the *sequence parameter set* (SPS) that contains important header information that applies to an entire series of coded pictures, and the *picture parameter set* (PPS) that contains header information that applies to one or more individual pictures within such a series. The *data partitioning* step, when used, formats the coded information such that the video data is more robust to data losses.

As in all prior ITU-T and ISO/IEC JTC1 video standards since H.261 in 1990, the VCL design of AVC follows the so-called block based hybrid video coding

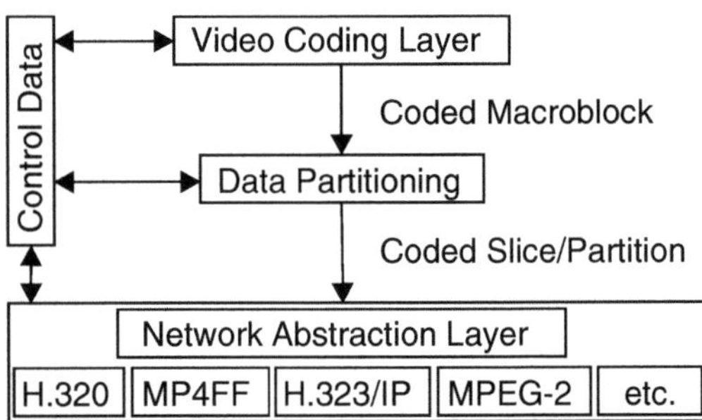

Fig. 3.1 AVC layer structure

approach (as depicted in Fig. 3.4 for the specific case of AVC). A *coded video sequence* (CVS) in AVC consists of an independently-coded sequence of *coded pictures* to which the same SPS header data applies. The VCL data for each picture is sent with a reference to its PPS header data, and the PPS data contains a reference to the SPS header data. In addition to motion-compensated inter-picture prediction to exploit temporal dependencies between different pictures that represent different time snapshots of the scene content, AVC also uses a more sophisticated intra-picture prediction to exploit dependencies between spatially-neighbouring blocks within the same picture. Spatial block transform coding is employed to exploit the remaining spatial statistical dependencies within the prediction residual signal for a block region (both for inter- and intra-picture prediction). So-called *macroblocks* of size 16×16, including all the associated multiple color components that represent these regions, are the basic building blocks for which the decoding process (including the signaling of the prediction references) is specified. In AVC, the macroblocks can be further sub-divided into smaller block-shaped units (with 4×4 being the smallest), e.g., for motion compensation (MC) and transform operations. *Slices* are sequences of macroblocks which are typically (except when using the *flexible macroblock ordering* feature further discussed below) processed in the order of a raster scan (i.e., starting at some position and scanning from left to right across the columns of its row, then the columns of the next row, etc.). A picture may be split into one or several slices as shown in Fig. 3.3. Slices are self-contained in the sense that given the active sequence and picture parameter sets, their syntax elements can be parsed from the bitstream and the values of the samples in the area of the picture that the slice represents can be correctly decoded and reconstructed.

The coding representation in AVC is primarily agnostic with respect to the original video characteristics, i.e., the underlying interlaced or progressive sampling and timing of the original captured pictures. Instead, its coding specifies a representation based on mathematical and geometric concepts, such as the structuring of data into rectangular regions of *frames* and *fields*, where each field consists of every-other line within a frame (either with *top field* or *bottom field* parity). A coded *picture* can represent either a complete frame or just a single field. The decoding process does not depend on the timing of the pictures, and pictures are transmitted and decoded in a particular order based on their encoded data dependencies rather than their intended display order.

For definition of a conforming decoder, it is not sufficient to just provide a description of the data decoding algorithm. It is also important in a real-time system to specify how (and when) bits are fed to a decoder and how the decoded pictures are removed from a decoder. By specifying input and output buffer models and developing an implementation-independent model of a receiver, the *hypothetical reference decoder* (HRD) achieves this, which is explored in further detail in Ref. [2]. An encoder is not allowed to create a bitstream that cannot be decoded by the HRD. Hence, in any receiver implementation, if the design adequately mimics the behavior of the HRD (adding some extra buffering to account for practical picture decoding speeds), it is guaranteed to be able to decode all conforming bitstreams. In AVC, the HRD specifies the operation of two buffers: (1) the *coded picture buffer* (CPB) and (2) the *decoded picture buffer* (DPB). The CPB models the arrival and removal timing of the coded bits. The HRD design is similar in spirit to buffer concepts in MPEG-2, but is more flexible in support of sending the video at a variety of bit rates without excessive delay. In AVC, multiple frames can be stored for use as references for decoding later pictures in bitstream order, and the reference frames can be located either in the past or future arbitrarily in display order. The decoder replicates the multi-picture buffer of the encoder in its DPB according to the reference picture buffering type and *memory management control operations* (MMCO) specified in the bitstream. Therefore, the HRD also specifies a model of the DPB to ensure that excessive memory capacity is not needed in a decoder to store the pictures used as references and to store pictures for reordering between bitstream order and display order.

In addition to basic coding tools, the AVC standard enables sending extra supplemental information along with the compressed video data. This often takes a form called *supplemental enhancement information* (SEI) or *video usability information* (VUI) in the standard. SEI data is specified in a backward-compatible way, so that as new types of supplemental information are specified, the new SEI can even be used with profiles of the standard that had previously been specified before that definition. Decoders simply ignore the SEI messages that they have not been designed to handle. Some examples of supplemental and auxiliary data are:

- Auxiliary pictures, which are extra monochrome pictures sent along with the main video stream, that can be used for such purposes as alpha blend compositing (specified as a different category of data than SEI).
- Film grain characteristics SEI, which allows a model of film grain statistics to be sent along with the video data, enabling an analysis-synthesis style of video enhancement wherein a synthesized film grain is generated as a post-process when decoding, rather than burdening the bitstream data with the representation of exact film grain during the encoding process.
- Frame packing arrangement SEI indicators to support 3D stereo video, which allow the encoder to identify the use of the video on stereoscopic displays, i.e., embedding left and right pictures into one single video frame with proper identification of how this is done and which pictures are intended for viewing by each of the viewer's eyes.
- Scalability and multi-view SEI indicators, which give additional hints to the decoder about best actions to take in cases where only partial information is available.
- Post-processing hint SEI indicators, which e.g. specify parameters of postprocessing filters which would lead to an improvement bringing the decoded picture closer to the original.

Reference software and a conformance testing set of example encoded bitstreams have been developed along with the standard. They can be used to facilitate the development of products that support the standard and to help provide a level of assurance that products actually conform to the standard [1].

3.3 Network Abstraction Layer

The NAL facilitates the ability to map the VCL data onto transport layers such as RTP/IP or MPEG-2 Part 1 systems (ISO/IEC 13818-1 and ITU-T Rec. H.222.0), or file formats such as the ISO base media file format. The full degree of customization of the video content to fit the needs of each particular application is outside the scope of the AVC standard, but the design of the NAL anticipates a variety of such mappings. Some key concepts of the NAL are *NAL units*, which are the basic packages of data to be conveyed by the bitstream, and the byte stream and packet format uses of NAL units, *parameter sets*, and *access units*. More detailed descriptions of these subjects, including error resilience aspects, are provided in Refs. [3, 4].

The coded video data is organized into NAL units, each of which is effectively a packet that contains an integer number of bytes. The first byte of each NAL unit is a header byte that contains an indication of the type of data in the NAL unit, and the remaining bytes contain payload data of the type indicated by the header. The NAL unit structure definition specifies a generic format for use in both packet-oriented and bitstream-oriented transport systems, and a series of NAL units generated by an encoder is referred to as a NAL unit stream.

For use in byte-oriented systems (e.g., MPEG-2 Part 1), each NAL unit is prefixed by a specific pattern of three bytes called a start code prefix. The start code prefix is simply two bytes that are equal to 0 followed by one byte equal to 1 (the hexadecimal pattern 0–000001). The boundaries of the NAL unit can then be identified by searching the coded data for the unique start code prefix pattern. *Emulation prevention bytes* are inserted as necessary within NAL units to guarantee that start code prefixes are unique identifiers of the start of each new NAL unit. A small amount of additional data (one extra byte per video picture, or optionally more at the encoder's discretion) is also added to support decoders that operate in systems using streams of bits without alignment to byte boundaries.

In packet-oriented systems (e.g., internet protocol/RTP systems), the coded data is carried in packets that are framed by the system transport protocol, and identification of the boundaries of NAL units within the packets can be established without the use of start code prefix patterns. However, the formatting of the data within the NAL units is the same, regardless of whether a byte stream format or packet format is in use to convey them.

NAL units are classified into VCL and non-VCL NAL units. The VCL NAL units contain the data that represents the values of the samples in the video pictures, and the non-VCL NAL units contain any associated additional information such as *parameter sets* (important header data that can apply to a large number of VCL NAL units) and *supplemental enhancement information* (timing information and other supplemental data that may enhance usability of the decoded video signal but are not necessary for decoding the values of the samples in the video pictures).

A parameter set contains highly-important information that is expected to rarely change, and the same parameter set can be used for the decoding of a large number of VCL NAL units. There are two types of parameter sets: *sequence parameter sets* (SPS)

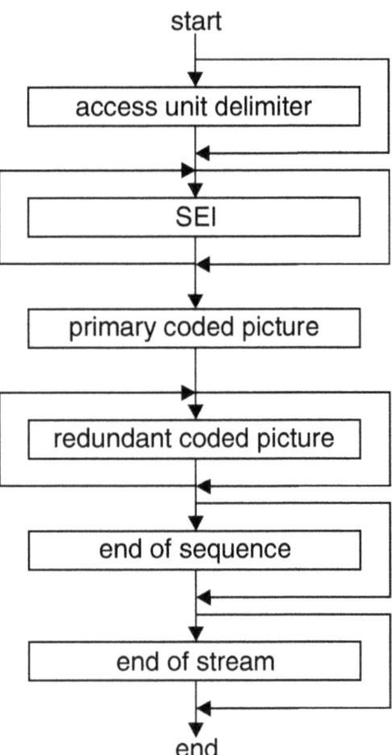

Fig. 3.2 Structure of access units

and *picture parameter sets* (PPS). This parameter-set mechanism decouples the transmission of infrequently changing information from the transmission of coded representations of the values of the samples in the video pictures. Each VCL NAL unit contains an identifier that refers to the content of the relevant PPS, and each PPS contains an identifier that refers to the content of the relevant SPS.

A set of NAL units in a specified form is referred to as an *access unit*. The decoding of each access unit results in one decoded picture. The format of an access unit is shown in Fig. 3.2. Each access unit contains a set of VCL NAL units that together compose a *primary coded picture*. It may also be prefixed with an *access unit delimiter* to aid in locating the start of the access unit. Some *supplemental enhancement information* containing data such as picture timing information may also precede the primary coded picture.

The primary coded picture consists of a set of VCL NAL units consisting of *slices* or *slice data partitions* that represent the samples of the video picture (Fig. 3.3).

A sequence of pictures that is independently decodable and uses a single SPS is referred to as a *coded video sequence* (CVS). If the coded picture is the last picture of a CVS, an *end of sequence* NAL unit may be present to indicate the end of the sequence; and if the coded picture is the last coded picture in the entire NAL unit stream, an *end of stream* NAL unit may be present to indicate that the stream is ending.

The series of access units for each CVS are sequential in the NAL unit stream. Each CVS can be decoded independently of any other CVS, given the necessary parameter set information, which may be conveyed either "in-band" within the video stream or "out-of-band" by some other mechanism. Each CVS starts with an *instantaneous decoding refresh* (IDR) access unit. An IDR access unit contains an

Fig. 3.3 Example of a slice structure (not using the FMO feature discussed below in Sect. 3.1.4.2)

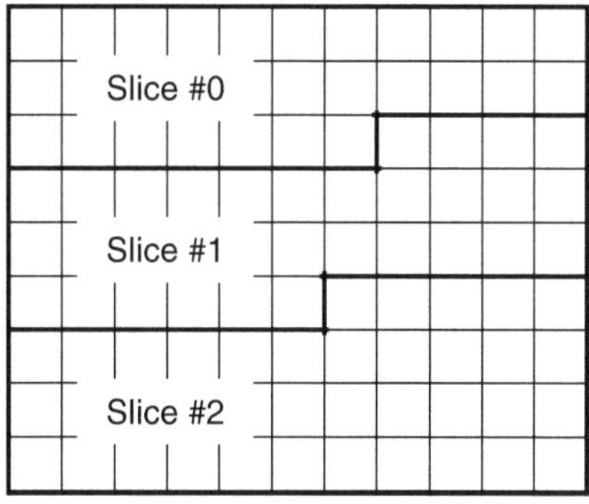

intra picture – a coded picture that can be decoded without decoding any previous pictures in the NAL unit stream, and the presence of an IDR access unit indicates that no subsequent picture in the stream will require reference to pictures prior to the intra picture it contains in order to be decoded. A NAL unit stream may contain one or more coded video sequences.

3.4 Coding Architecture and Tools of the VCL

3.4.1 Structure of Encoder and Decoder Processing

A block diagram of the hybrid video coding structure as it is applied in AVC is shown in Fig. 3.4. The main decoder operations are

- Bitstream parsing and entropy decoding,
- Reconstruction of transform coefficients and inverse transform to generate the residual signal,
- Generation of the prediction signal from previous decoded pictures based on decoded side information such as motion vectors, block modes, etc.,
- Superposition of the residual signal with the prediction signal to generate the decoded representation.
- Filtering the final result with a deblocking filter to remove block-shaped coding artifacts and improve the quality of the picture for use in subsequent prediction operations.

The encoder runs the same prediction loop as the decoder to generate the same prediction signal and subtracts it from the original picture to generate the residual. The most difficult task of the encoder is to determine its choices for prediction (such as motion vectors, block sizes, and inter/intra prediction modes) and residual difference coding, for which the encoder tries to use as few bits as possible after entropy coding to obtain adequate decoded quality. The way an encoder makes these decisions is left entirely outside the scope of the standard, so different encoder products

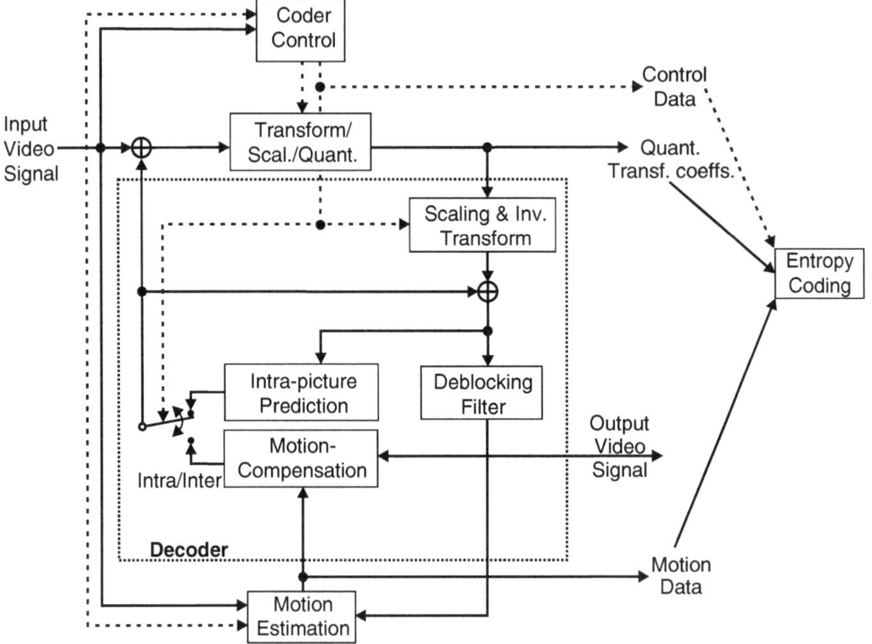

Fig. 3.4 Typical structure of an AVC video encoder

may have quite different compression capability and also may have quite different computational resource consumption.

The VCL of AVC has many design features in each of the elements above, which provides an encoder with many more degrees of freedom than previous standards. The most important elements of the design are described in the subsequent subsections.

3.4.2 Slices, Slice Types, Slice Groups, Slice Data Partitioning

Each slice can be coded using different coding types as follows, where the first three are very similar to those in previous standards (with the exception of the use of reference pictures in various video sequence prediction structures as described further below):

- **I slice**: A slice in which all macroblocks of the slice are coded using intra-picture prediction.
- **P slice**: In addition to the coding types of the I slice, some macroblocks of the P slice can also be coded using inter-picture prediction with at most *one* motion-compensated prediction signal per prediction block.

- **B slice**: In addition to the coding types available in a P slice, some macroblocks of a B slice can also be coded using inter-picture prediction with *two* motion-compensated prediction signals per prediction block.
- **SP slice**: A so-called switching P slice that is coded such that efficient switching between different pre-coded pictures becomes possible.
- **SI slice**: A so-called switching I slice that allows an exact match of a macroblock in an SP slice for random access and error recovery purposes.

Each slice consists of a set of macroblocks in raster scan order, except when the feature known informally as *flexible macroblock ordering* (FMO) is used. In the FMO case, the macroblocks of the picture are first partitioned into groups called *slice groups*, and then the macroblocks of a slice consists of a set of macroblocks in raster scan order that are within the same slice group. The segmentation of the picture into slice groups is performed using a map called the *macroblock-to-slice-group map*. It can be used to separate some area for separate coding as a "region of interest" for special treatment, or for other purposes such as enhanced robustness to data losses.

Each slice within each picture is independently decodable (given the parameter sets and previously decoded pictures on which it depends), so that it is possible for the slices of each picture to be decoded by separate parallel processes or to be decoded in any order relative to each other – and to ensure that even if some slices are lost by the communication channel, the others can still be decoded. A feature known as *arbitrary slice ordering* (ASO) can even allow the slices to be sent in any order within the access unit for the coded picture. Further enhanced loss robustness can be achieved by using a feature known as *redundant pictures* to send extra slices representing (ordinarily reduced-quality) redundant representations of some pictures or picture regions.

When applied, the *data partitioning* feature can segment the coded video data into different categories in a way that can be used to enhance loss robustness. When data partitioning is applied, the data for each slice is separated into three *slice data partitions*: one for prediction and segmentation information, one for the transform coefficients of intra-picture coded regions, and a third for the transform coefficients of inter-picture predicted regions.

3.4.3 Video Sequence Prediction Structures and Decoded Picture Buffering

In terms of video sequence prediction structures, AVC is much more flexible than previous standards due to its sophisticated reference decoded picture buffer management and its ability to signal the reference picture(s) to be used for each respective macroblock. (A form of this feature appeared first in H.263 Annex U, but reached substantially greater maturity in AVC.) This allows for structures that substantially improve the compression gain such as multiple-reference picture prediction and hierarchical prediction structures of B slices (Fig. 3.5). Rather than only

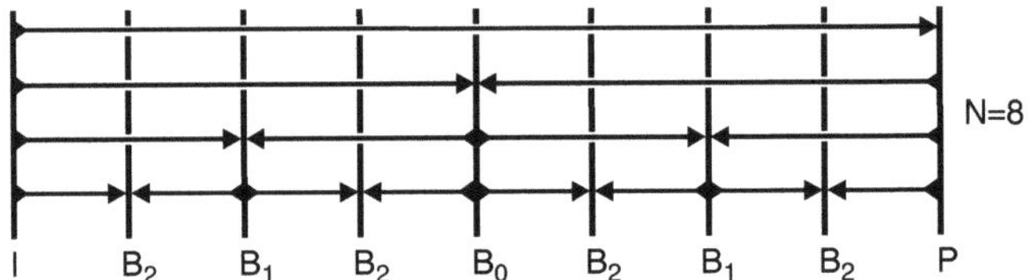

Fig. 3.5 An example hierarchical B slice picture prediction structure

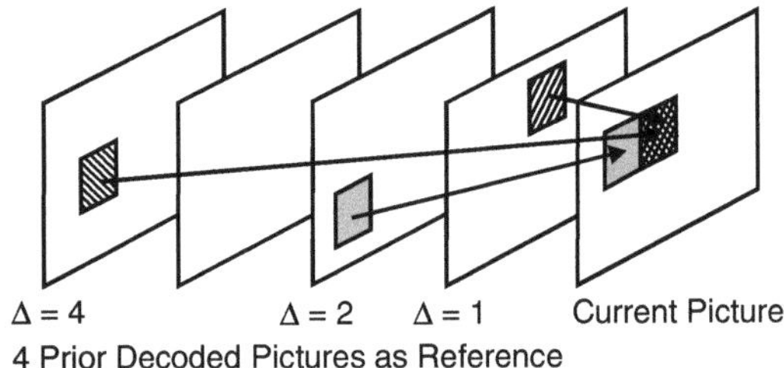

Fig. 3.6 Multi-frame motion compensation. In addition to the motion vectors, the reference picture indexes (Δ) are transmitted

having a single picture that can be used as a reference for decoding a P slice, a list of potential reference pictures is formed, and the encoder can indicate which of these pictures is used for prediction of any particular region (Fig. 3.6). The substantial difference between B and P slices is that B slices are coded in a manner in which some macroblocks or blocks may use a (possibly weighted) average of two distinct motion-compensated prediction values for building the prediction signal. To support the generation of these two prediction values, B slices utilize two distinct lists of reference pictures, which are referred to as the "list 0" and "list 1" reference picture lists, while P slices use only one prediction signal for each region, and have only one reference picture list – list 0. Each list is constructed by identifying the length of the list and identifying which picture in the DPB is placed at each position in the list. Which pictures are actually located in each reference picture list is an issue of the multi-picture buffer control established by the encoder.

In B slices, four different types of inter-picture prediction are supported: list 0, list 1, *bi-predictive*, and *direct* prediction. For the list 0 and list 1 prediction types, a single prediction signal is used, based on a picture in the corresponding list, and in the bi-predictive mode, the prediction signal is formed by a weighted average of motion-compensated list 0 and list 1 prediction signals. The direct prediction mode infers the type of prediction for a region, based on previously transmitted syntax elements for neighbouring regions. Once this inference rule operates, direct prediction becomes effectively the same as list 0 prediction, list 1 prediction, or bi-prediction.

Multiple reference picture prediction allows the encoder to define references for the prediction of macroblock or sub-macroblock regions from any of the previously decoded pictures in the list(s); this is completely agnostic of the actual display order or timing. In the case of the bi-predictive mode, it is not prescribed that one of the

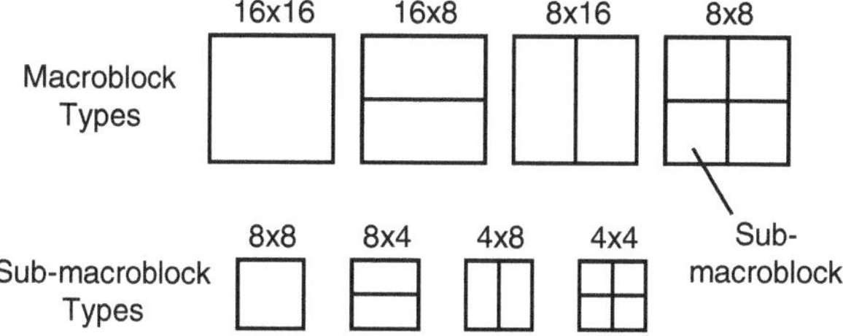

Fig. 3.7 Partitioning of a macroblock (*top*) and a sub-macroblock (*bottom*) for motion-compensated prediction

pictures has to be a previous frame in display order and the other has to be a future frame in display order (as was the case in prior designs such as MPEG-1 and MPEG-2 video), and single list prediction in P slices is also not restricted to refer only to pictures that precede the current picture in display order (as was also the case in such prior designs). There is also no prohibition against having B slices used as references for the prediction of other pictures (as was also the case in such prior designs). The indication of whether any picture may be used as a reference picture for the subsequent decoding of other pictures is completely decoupled from the type of slices (I, B, or P) used for its encoding, and different slice types can be used for different slices within the same picture.

3.4.4 Macroblock and Sub-Macroblock Structures

Below the 16×16 macroblock level, the AVC standard supports more flexibility in the selection of motion compensation block sizes and shapes than any previous standard. First, a subdivision into 8×8, 8×16 or 16×8 prediction regions can be made, (see macroblock types in Fig. 3.7). In case of a sub-macroblock of size 8×8, further subdivision into 4×4, 4×8 and 8×4 block sizes is possible, (see sub-macroblock types in Fig. 3.7), which allows much higher accuracy with regard to discontinuous motion. Whereas prior video coding standards used only a transform block size of 8×8, the initial AVC design was based primarily on a 4×4 transform, and an 8×8 transform was also included in the newer profiles (with an ability for the encoder to adaptively switch the transform block size from macroblock-to-macroblock).

3.4.5 Interframe Prediction

Due to the higher flexibility in the selection of motion compensation block sizes and shapes, as well as the concept of multiple-reference pictures, the quality of motion compensated prediction is significantly increased. Furthermore, quarter-sample motion vector accuracy is used, where the complexity of the interpolation processing was reduced compared to the design of the Advanced Simple profile of the MPEG-4 Visual (MPEG-4 Part 2) standard (which was the only substantial use of quarter-sample motion vector accuracy in prior standards). Weighted prediction is new in AVC, which allows the motion-compensated prediction signal(s) to be weighted and offset by amounts specified by the encoder. This can sometimes dramatically improve coding efficiency – e.g. for scenes containing fades.

3.4.6 Intraframe Prediction

For intra coding modes, AVC specifies the prediction of samples in a given block in a manner conducted in the spatial domain, which means that a prediction residual signal is transformed and encoded in intra coding as well as in inter-picture predicted coding.

Spatially neighbouring samples of a given block that have already been transmitted and decoded are used as a reference for spatial prediction of a given block's samples. This prediction process exploits dependencies across transform block boundaries and compensates for some of the disadvantages of using small transform block sizes. Compared to that, previous video coding standards performed some prediction across block boundaries in the transform domain, which is substantially less effective. The intra prediction types are distinguished by their underlying luma prediction block sizes of 4×4, 8×8 (some profiles only), and 16×16. The number of encoder-selectable prediction modes in each intra coding type is either four (for chroma and 16×16 luma blocks) or nine (for 4×4 and 8×8 luma blocks). Prediction is performed from the previous neighbouring blocks' closest boundary samples, where the modes are either prescribing a certain direction of propagation into the current block, or using the average ("DC" prediction) for prediction of the entire block as a flat region (Fig. 3.8). The intra prediction process for chroma samples operates in an analogous fashion but with a prediction block size equal to the block size of the entire macroblock's chroma arrays.

In order to provide encoders with the ability to tune the degree of loss robustness in the design, encoders may either select to allow intra prediction from any region of the picture, or only from those regions that were not coded using inter-picture prediction from other previously decoded pictures. The latter case is referred to as *constrained intra prediction*, and it can help prevent the propagation of picture corruptions that can be caused by the operation of the inter-picture prediction processing.

3.4.7 Motion Vector Prediction

Due to the higher accuracy in the motion vector field (both in terms of the supported prediction block sizes and the use of quarter-pel motion vector precision), the bit

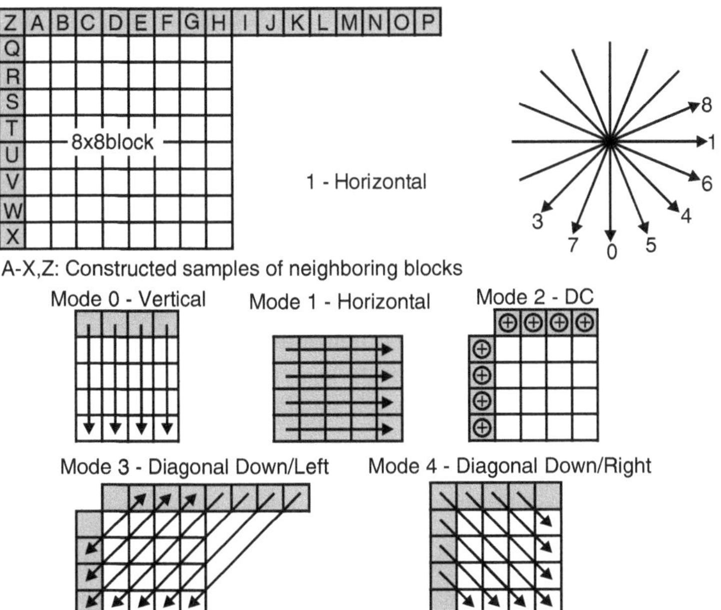

Fig. 3.8 *Top*: Samples used for 8×8 spatial luma intra prediction (*left*) and directions of 4×4 and 8×8 spatial luma intra prediction modes 0, 1, and 3–8 (*right*). *Bottom*: Five of the nine Intra 4×4 prediction modes

rate needed for motion vector encoding is a key component of the rate-distortion performance. Therefore, the median value of the motion vectors of neighbouring blocks is used for prediction of the current motion vector, and the motion vector prediction residual is entropy encoded. Furthermore, hierarchical prediction is applied below the macroblock level. Whereas some prior standards allowed for "skipping" of a given area of a predictively-coded only in areas of static (zero motion) scene content, the AVC design infers motion in "skipped" areas. For B slices, AVC also includes the enhanced motion inference method known as "direct" motion compensation, which further improves upon the "direct" prediction designs as found in the prior H.263 and MPEG-4 Visual standards.

3.4.8 *Transform and Quantization*

AVC specifies 2D integer transforms primarily with two block sizes: 4×4 and 8×8. In the profiles related to the first release of AVC (see Sect. 3.1.6.1), primarily only a 4×4 integer transform is applied to the prediction residual signal – for both the luma and chroma components. The small block size of this transform allows the encoder to represent signals in a more locally-adaptive fashion, which reduces artifacts known colloquially as "ringing" or "mosquitos". The smaller block size also corresponds to the size of the smallest prediction regions (both for motion compensated and intra prediction). In addition, a hierarchical block transform (Hadamard/Haar type) is applied to the chroma and to a 4×4 array of the DC coefficients of a prior set of 4×4 transforms in some luma regions for intra coding, effectively extending the block size to 16×16 and giving higher compression in the case of highly correlated signals. The 4×4 integer transform is strikingly simple to compute – much easier to compute than the discrete cosine transform (DCT) specifications in prior standards – which had actually been specified in terms of infinite-precision irrational trigonometric function values to which each decoder designer was responsible for finding a reasonably close approximation within a tolerated specified precision. By specifying the transform using exact integer values, the computational requirements of the transform were reduced while actually improving the quality of the result by allowing the standard to require exact-match decoded results.

For high-fidelity video, however, the preservation of smoothness and texture can generally benefit from a representation with longer basis functions than a 4×4 transform can provide. This was achieved by making use of an 8×8 integer transform in profiles starting from the FRExt amendment as an additional transform type for coding the luma residual signal. This 8×8 block transform can be viewed as an approximation of the 2D 8×8 DCT in integer arithmetic. In fact, all integer transforms in AVC as well as their corresponding inverse transforms can be implemented using only shift and add operations in $(8+b)$-bit arithmetic precision for processing of b-bit input video, which keeps the dynamic range within the important 16-bit bounds for the most common $(b=8)$ input video case. The encoder has the choice between using the 4×4 or 8×8 transform on a macroblock-by-macroblock basis. This adaptive choice is coupled together with related parts of the decoding process – for example, by disallowing use of the 8×8 transform when the prediction block size is smaller than 8×8.

For the quantization of transform coefficients, AVC uses uniform-reconstruction quantizers (URQs). One of 52 quantizer step size scaling factors (plus six for each additional bit of source video bit depth beyond 8 bits) is selected for each macroblock by a *quantization parameter* (QP). The scaling operations are arranged so that there is a doubling in quantization step size for each increment of six in the value of QP. Its logarithmic style of quantization step size control makes step size control easier for encoder bit rate control purposes, and eases the computational burden as well due to the way the step size is used in the inverse quantization process. In addition to the basic step-size control, the newer profiles starting from those of the FRExt amendment also support encoder-specified scaling matrices for a perceptually tuned, frequency-dependent quantization (a capability similar to that found in MPEG-2 video). The quantized transform coefficients of a block generally are scanned in a zig-zag fashion and further processed by using the entropy coding methods as described below. Some alternative scan methods other than the ordinary zig-zag scan are employed for the cases of intra predictive coding and inter field-structured coding.

3.4.9 Lossless Macroblock Representations

AVC supports a special type of macroblock called the PCM (literally *pulse-code modulation*, a term that has lost its original historic meaning) macroblock in which the values of the samples in the macroblock are simply sent directly, without using a transform or quantization. This allows the encoder to represent selected regions perfectly, and allows the standard to specify a strict limit on the number of encoded bits that may be used to represent any individual macroblock (by setting the maximum number of encoded bits to approximately the value used by a PCM macroblock).

It also includes a transform-bypass lossless mode which uses prediction and entropy coding for encoding sample values. When this mode is enabled, the meaning of the smallest selectable value of the quantization parameter is redefined to invoke the lossless coding operation. This is a very efficient lossless video coder due to its ability to be used with inter-picture prediction. No other standard supports lossless video coding with inter-picture prediction.

3.4.10 Entropy Coding

AVC uses much more sophisticated entropy coding than previous video coding standards; in particular it uses adaptive and context-dependent coding for better exploitation of source statistics and coherences. By this, it becomes possible e.g. to define special modes and code the related side information efficiently, even in cases where such special modes may not be used frequently by an encoder.

Firstly, many syntax elements are coded using the same highly-structured infinite-extent variable-length code (VLC), called a zero-order exponential-Golomb code. A few syntax elements are also coded using simple fixed-length code representations. For the remaining syntax elements, two different entropy coding mechanisms are defined, one of which is *context-adaptive VLC* (CAVLC), the other *context-adaptive binary arithmetic coding* (CABAC), each of them used in dedicated profiles. CABAC is also employed to perform further compression of the other syntax elements.

For lower-complexity applications, the exponential-Golomb code is used for nearly all syntax elements except those of quantized transform coefficients, for which CAVLC is employed. When using CAVLC, the encoder switches between different VLC tables for various syntax elements, depending on the values of the previously-transmitted syntax elements in the same slice. Since the VLC tables are designed to match the conditional probabilities of the context, the entropy coding performance is improved in comparison to schemes that do not use context-based adaptivity.

The entropy coding performance is further improved when CABAC is used. As depicted in Fig. 3.9, the CABAC design is based on three components: binarization, context modeling, and binary arithmetic coding. Binarization enables efficient

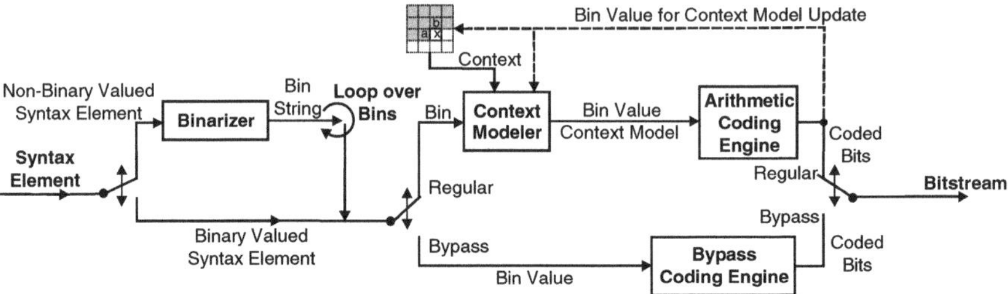

Fig. 3.9 CABAC block diagram

binary arithmetic coding by mapping non-binary syntax elements to sequences of bits referred to as bin strings. The bins of a bin string can each either be processed in an arithmetic coding mode or in a *bypass* mode. The latter is a simplified coding mode that is chosen for selected bins such as sign information or lesser-significance bins, in order to speed up the overall decoding (and encoding) processes. The arithmetic coding mode provides the largest compression benefit, where a bin may be context-modeled and subsequently be arithmetic encoded. As a design decision, in most cases only the most probable bin value of a syntax element is supplied with external context modeling, which is based on previously decoded (encoded) bins. The compression performance of the arithmetic-coded bins is optimized by adaptive estimation of the corresponding (context-conditional) probability distributions. The probability estimation and the actual binary arithmetic coding are conducted using a multiplication-free method that enables efficient implementations in hardware and software. Compared to CAVLC, CABAC can typically provide reductions in bit rate of 10–20% for the same objective video quality when coding standard-definition (e.g., 720×576) or high-definition (e.g., 1,920×1,080) video material.

3.4.11 *Interlaced Video Handling*

The principles for special treatment of field-captured video are basically conceptually similar in AVC as they were in MPEG-2 (ISO/IEC 13818-2 and ITU-TH.262). To provide high coding efficiency, the AVC design allows encoders to make any of the following decisions when coding two fields of such video.

- To combine the two fields together and to code them as one single coded frame (frame mode).
- To not combine the two fields and to code them as separate coded fields (field mode).
- To combine the two fields together and compress them as a single frame, but when coding the frame to code some regions as a combined frame area and others as two distinct field regions.

The choice between the three options can be made adaptively for each frame in a coded video sequence. The choice between the first two options is referred to as picture-adaptive frame/field (PAFF) coding. When a frame is coded as two separate fields, each field is partitioned into macroblocks and is coded in a manner very similar to a frame, with the following main exceptions:

- Motion compensation uses reference fields rather than reference frames;
- The zig-zag scan of transform coefficients is somewhat different (as vertical high frequencies are more dominant in field-structured coding);
- The strongest deblocking filter strength is not used for filtering horizontal edges of macroblocks in fields, because the field rows are spatially twice as far apart as frame rows and the length of the filter thus covers a larger spatial area.

If a frame consists of mixed regions where some regions are moving and others are not, it is typically more efficient to code the non-moving regions in frame mode and the moving regions in the field mode. Therefore, a frame/field encoding decision can also be made separately for each vertical pair of macroblocks (each 16×32 luma region) in a frame. This coding option is referred to as macroblock-adaptive frame/field (MBAFF) coding. Unlike in MPEG-2, the frame/field decision is made at this macroblock pair level rather than within the macroblock level, so that the frame/field coding decision for each such region can be coded more efficiently and so that the coding design has no need to handle the half-macroblocks that arise in the corresponding case in MPEG-2. Each macroblock of a field macroblock pair is processed very similarly to a macroblock within a field in PAFF coding. However, since a mixture of field and frame macroblock pairs may occur within an MBAFF frame, the methods that are used for zig-zag scanning, prediction of motion vectors, prediction of intra prediction modes, intra-frame sample prediction, deblocking filtering, and context modeling in entropy coding are modified to account for this mixture. Another important distinction between MBAFF and PAFF is that in MBAFF, one field cannot use the macroblocks in the other field of the same frame as a reference for motion prediction (in order to simplify the design and implementation). Thus, sometimes PAFF coding can be more efficient than MBAFF coding (particularly in the case of rapid global motion, scene change, or intra picture refresh), although the reverse is usually true.

3.4.12 Deblocking Filtering

Due to coarse quantization at low bit rates, block-based coding can often result in visually noticeable discontinuities along the block boundaries. Furthermore, due to the properties of block predictors and DCT-like basis functions, coding errors tend to become maximized near the block edges. If no further provision is made to deal with this, these artificial discontinuities may also propagate into the interior of blocks in subsequent coded pictures by means of the motion-compensated prediction process. This could become even more severe in the AVC design since it supports small block sizes such as 4×4 in the motion compensation and transform stages. For that purpose, AVC defines a deblocking filter which operates within the predictive coding loop, and thus constitutes a required component of the decoding process. The filtering process exhibits a high degree of content-adaptivity on different levels, from the slice level along the edge level down to the level of individual samples.

The basic idea is that if a relatively large absolute difference between samples near a block edge is measured, it is quite likely a blocking artifact and should therefore be reduced. However, if the magnitude of that difference is so large that it cannot be due to the coarseness of the quantization step size used in the encoding, the edge is more likely to reflect the actual behavior of the source picture and should not be smoothed over. The blockiness is reduced by the deblocking filter, while the sharpness of the content is basically unchanged. Parameters to adapt the strength of the filtering are derived from the sample differences around the block boundaries, the coarseness of the quantization parameter, the number of coefficients retained, the differences of motion vectors of adjacent blocks, and differences in the coding modes of adjacent blocks such as intra/inter modes or reference frame indexes. Moreover, the strength of the filter can be adjusted by syntax sent by the encoder, so that encoder designers can further optimize its behavior. The filter itself is a linear 1D filter with directional orientation perpendicular to the block boundary and impulse response lengths between three and five taps depending on the filter strength. The amount of modification induced by the 1D filter is then constrained within adaptively determined limits. A detailed description can be found in Ref. [5]. As a result, the blockiness is reduced without much affecting the sharpness of the content. Consequently, the subjective quality is significantly improved, and some objective (PSNR) gain can also be achieved – particularly at low rates.

3.4.13 Robustness to Data Loss and Network Characteristics

The structure of AVC is designed with a variety of features to make it robust to data losses. These include the network abstraction layer structure, parameter sets and the ways they can be carried and repeated to enhance robustness, independently-decodable coded video sequences, I slices, independence of the decoding process of each slice within each picture, constrained intra prediction, data partitioning, flexible macroblock ordering, arbitrary slice ordering, and redundant pictures. However, loss robustness is also highly dependent on important issues that are outside the scope of the video coding standard itself, such as the many decisions made by an encoder, the system environment in which the video bitstream is being conveyed (and the loss protection it may provide), and the way that a decoder detects and conceals data losses when they occur. Such aspects are further discussed in Refs. [3, 4].

3.5 Extended Functionality

3.5.1 Scalable Video Coding (SVC)

Scalability provides a feature where parts of the bitstream can be discarded, however decoding is still possible at lower quality or resolution. The SVC approach of AVC is based on the "layered coding" principle, which encodes differential information between layers of different quality or resolution. Advantages of scalability are

- If a video signal must be provided at various resolutions (e.g. SD and different HD video formats in broadcast and internet streaming), SVC allows more efficient transmission than by independent streams (simulcast).
- Post-encoding modification of bitstream sizes is simplified (e.g. down-sizing bitstreams to clear memory space); it is not necessary to apply transcoding.
- Unicast transmission services with uncertainties regarding channel conditions (throughput, errors) or client device types (supported spatio-temporal resolution by decoder, display and power) benefit from the fact that bitstreams can easily be adapted and if necessary protected anywhere in the transmission chain. This is even more true for multicast or broadcast transmission services where the diversity of uncertainties is even larger than in unicast transmission.

The most important properties of SVC are as follows:

- The core of information is a base layer which is compatible with the existing Constrained Baseline or High profiles (for the aforementioned SVC profiles, respectively). Base layer packets are encapsulated in NAL unit types which are known to existing single-layer decoders and can be extracted by those, whereas the scalable enhancement information is conveyed in SVC specific (and previously unspecified) NAL unit types that are ignored by prior version decoders, and as such the enhancement data can only be decoded by SVC capable decoders.
- Within the enhancement layers, a unique dependency structure is defined which is signaled in the NAL unit header. By this, it can be derived which lower layers are necessary for decoding the respective packet, and devices such as switches or any kind of proxy or network nodes with sufficient knowledge about the header structure can decide whether to convey or discard a given packet. IETF has also defined an RTP payload format for SVC [6] which adopts the same concept by copying the information of the NAL unit header into the RTP header.
- The usual entropy coding methods of AVC (CABAC and CAVLC) are extended by a few additional context models (e.g. contexts across layers).

The main properties that guarantee efficiency in scalability are:

- Temporal scalability is most efficient by usage of hierarchical frame prediction structures (see examples of a B frame pyramid and a P frame pyramid with four temporal layers each in Fig. 3.10). As a general rule, layers with higher temporal resolution shall not be used for prediction of layers with lower temporal resolution.

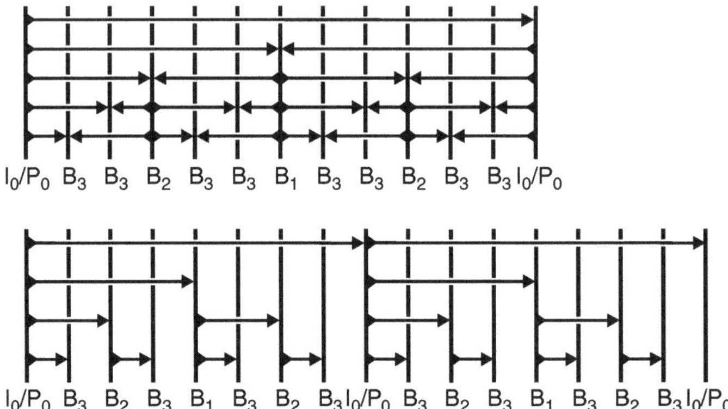

Fig. 3.10 Hierarchical frame prediction structures for temporal scalability. *Top*: Hierarchical B structure with 30/10/5/2.5 fps. *Bottom*: Hierarchical P for low delay encoding/decoding with 60/30/15/7.5 fps

Furthermore, pictures at the lowest layer of the hierarchy (I and P in Fig. 3.10) are denoted as key pictures and play a role of re-synchronization points. In principle, temporal scalability was already supported by non-scalable AVC through multiple reference picture syntax, when *reference picture list reordering* (RPLR) and *memory management control operations* (MMCO) are appropriately used. However, SVC provides enhanced scalability signaling support by its NAL unit header extension or separate SVC NAL units succeeding AVC NAL units.

- For spatial and SNR scalability, relevant information from lower layers is utilized as far as possible, including motion and mode information. To perform prediction from a lower to a higher layer, this is implemented as an additional prediction mode, whereas all other modes (e.g. motion compensation, intra prediction) are typically still available for that layer. Only blocks decoded in intra mode, or the residual signal of motion compensated blocks can be used for prediction across layers. This avoids full reconstruction of motion compensated blocks of all lower layers, and therefore multiple motion compensation loops at the decoder end are avoided.
- A typical configuration of spatial scalability is shown in Fig. 3.11. Here, layered coding results in an over-sampled pyramid. Motion compensation (MC) prediction structures of all layers are aligned. Prediction of partitioning and motion information in layer $n+1$ is performed by upsampling the corresponding information from layer n (e.g. enlarge block structures or motion vectors by factor of 2) and use them as switchable predictors. Likewise, intra-coded macroblocks and MC residual data are upsampled and used as optional additional predictors in the corresponding areas of layer n.
- For spatial scalability, often dyadic (powers of two) factors are used in terms of frame sizes of the layers; non-dyadic factors are also possible, e.g. supporting scalability between SDTV and 720p resolution. Scalability between interlaced and progressive is also supported in principle. Compared to simulcast, the efficiency

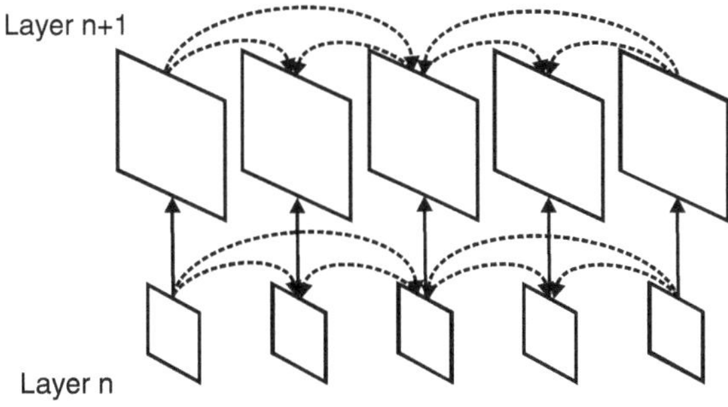

Fig. 3.11 Two layers of spatial scalability

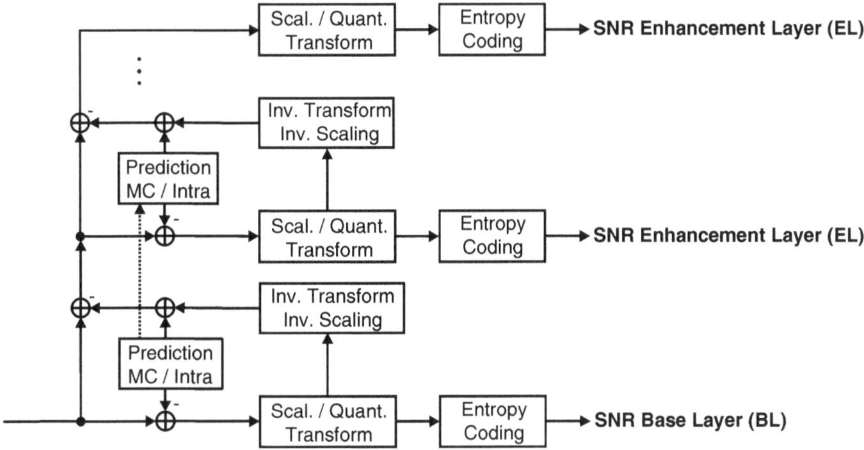

Fig. 3.12 Encoder operation for multiple SNR enhancement layers and multiple MC loops

of SVC is typically increased if it is necessary to provide two spatial resolutions which are relatively close to each other.

– SNR Scalability is implemented with two basic operation principles: Coarse-granular SNR scalability (CGS) and medium-granular SNR scalability (MGS). CGS is designed to support a few selected points with a typical minimum of 50% bit-rate increase from one layer to the next. It works like spatial scalability without spatial resolution change between the layers, only quantizer fidelity is changed. All enhancement layer NAL units should be available to guarantee the intended quality. In CGS, SVC bitstreams can be converted to single-layer AVC bitstreams by a method called *bitstream rewriting*. An encoder structure with refinement of motion vectors from one layer to the next is shown in Fig. 3.12. Even though multiple MC loops are run at the encoder, at the decoder only one IDCT and one MC loop are necessary (corresponding to the encoder loop of highest available layer). MGS is a combination of simple CGS syntax with a more advanced motion-compensated prediction structure, allowing the use of adaptive prediction from either base layer or enhancement layer, and assigning additional levels to sub-groups of transform coefficients. MGS NAL units are partially discardable, such that by principle layers between two CGS layers can be additionally defined.

By the "key picture" concept, an adjustable trade-off between coding efficiency and drift is achieved. Even though the decoder performs single-loop operations, multiple closed encoder loops run at different rate points in the overall layered structure (which may consist of various combinations of temporal, spatial and SNR scalability). Drift-free operation is guaranteed at those rate points where encoder loops are operated such that stability is achieved over a broad range of rates.

In terms of results, compared to single-layer coding of highest resolution standalone, SVC typically requires 5–10% higher rate (depending on sequence characteristics and with optimized encoder configuration). Likewise, the SVC performance is significantly better than simulcast (transmitting the bitstreams relating to different resolutions in parallel). SVC also performs significantly better and is much more flexible in terms of number of layers and combination of scalable modes than any scalable version of MPEG-2 and MPEG-4 Part 2. This is mainly due to the following reasons:

- SVC uses layered coding (bottom-up prediction) not just for the image information, but for any kind of information such as motion vectors, modes, residuals, etc.
- SVC can make extensive use of hierarchical frame structures as provided in AVC, which per se is more robust in case of partial data availability.
- SVC allows flexible control of the drift by running an encoder with multiple points of closed prediction loop.

At the same time, the decoder is less complex because it can be run with a single loop. The SVC extensions are further discussed in Ref. [7].

3.5.2 Multiview Video Coding (MVC)

In its basic design concepts, MVC is an extension of the inter-view prediction principle similar as implemented in the MPEG-2 multiview profile (assuming two or multiple cameras shooting the same scene). Compared to MPEG-2, MVC benefits from the more flexible multiple reference picture management capabilities in AVC. The key elements of the MVC design are:

- No changes to lower-level AVC syntax (slice and lower); core decoding modules are not aware of whether a reference picture is a time reference or a view reference.
- Small design-compatible changes to high-level syntax, e.g., to specify view dependency, random access points.
- A base view is easily extracted from the video bitstream (identified by NAL unit type byte); non-MVC decoders simply ignore the enhancement data and only decode the base view.
- The syntax allows decoded pictures from different views to be inserted in, and to be removed from the reference picture buffer.

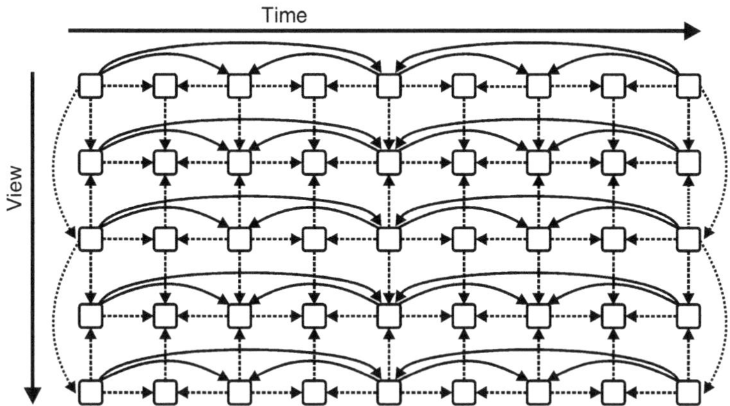

Fig. 3.13 Hierarchical prediction across dimensions of time and view in MVC

An example of a prediction structure for the case of five camera views is shown in Fig. 3.13, with pictures of the base view in the uppermost row. Inter-view prediction is generally limited to use references only from the same time instance.

Data rate savings compared to simulcast are typically 20–30% in mean square error terms for stereo, and can go to 50% and higher for multi-camera captures. In terms of perceptual quality, it has been reported that the benefits are greater. The compression advantage however highly depends on the amount of redundancy between the views (mainly depending on camera baseline distance) and the amount of temporal change in the scene. It can be concluded that the effect of utilizing inter-view redundancy is significantly larger than in MPEG-2 Multiview profile, which again comes mainly due to the fact that AVC allows a much higher degree of flexibility in the definition of prediction references. One emerging application of MVC is disc storage for stereo HD video, and it has been adopted as the technology for 3D Blu-ray disc encoding. These extensions and other types of 3D stereoscopic video support using AVC are further discussed in Ref. [8].

3.6 Profiles and Levels

Profiles and levels specify conformance points. These conformance points are designed to facilitate interoperability between various applications of the standard that have similar functional requirements. A *profile* defines a set of coding tools or algorithms that can be used in generating a conforming bitstream, whereas a *level* places constraints on certain key parameters of the bitstream. As of mid-2011, there will be 18 profiles defined in the standard, which are described in further detail below. All decoders conforming to a specific profile must support all features in that profile. Encoders are not required to make use of any particular set of features supported in a profile but have to provide conforming bitstreams, i.e., bitstreams that can be decoded by conforming decoders. The same set of level definitions is used with all profiles, but individual implementations may support a different level for each supported profile. There are 17 levels defined, specifying upper limits for the picture size (in macroblocks) ranging from QCIF all the way to common 4 K×2 K formats, decoder-processing rate (in macroblocks per second) ranging from 250 k pixels/s to 250 M pixels/s, video bit rate ranging from 64 kbps to 240 Mbps, as well as suitable limits of the picture buffer sizes.

The various profiles with their associated coding tools have been defined in several phases and are briefly described below.

3.6.1 Profiles Based on Features in the First Version (4 Profiles)

The first version defined the *Baseline*, *Main*, and *Extended Profiles*. Another profile called the *Constrained Baseline Profile* was also later defined that uses only features of the original version of the standard. In terms of tools, none of these supports the 8×8 transform, predictive lossless coding mode, bit depths of more than 8 bits per sample, color sampling other than 4:2:0, bitstream scalability features (other than some forms of temporal scalability) or multi-view captures.

These profiles support all coding features except those in the following feature sets:

- *Set A*: ASO, FMO, and redundant pictures.
- *Set B*: B slices, weighted prediction, field coding, and picture-adaptive and macroblock-adaptive switching between frame and field coding.
- *Set C*: CABAC entropy coding.
- *Set D*: SP/SI slices and slice data partitioning.

These sets of feature capabilities are allocated to the three original profiles as follows:

- *Baseline profile* (BP) supports the Set A features, which provide it with substantial added robustness to data losses and flexibility for packet network transmission order.
- *Main Profile* (MP) supports Set B and Set C in order to provide it with enhanced compression capability, but does not include support for Set A (to avoid the implementation work needed for those features, which were considered less necessary for entertainment-quality applications operating on high-reliability channels). Since the Main Profile does not support the FMO, ASO, and redundant pictures features that are supported by the Baseline profile, only a subset of the coded video sequences that are decodable by a Baseline profile decoder can be decoded by a Main Profile decoder.
- *Extended Profile* (EP) supports all features of the Baseline Profile, and the Set A, Set B, and Set D features on top of Baseline Profile, but it does not support Set C (CABAC). It was designed primarily for environments in which flexibility and loss robustness are especially important but in which implementation of CABAC was considered difficult (e.g., for software-only decoders).

- *Constrained Baseline Profile* (CBP) was defined in 2009 to mitigate the lack of "nesting" in the original profile structure. In the original standard, some flags were defined in the sequence parameter set to indicate which decoder profiles could decode a coded video sequence, but there was no common "core" profile of features that could be decoded by all three of the original decoder profiles in the first version. This "hole" in the standard was then filled by the additional profile that leaves out all four of the feature sets that are not supported in the other three profiles. Any decoder capable of decoding bitstreams of the Baseline, Main, or Extended Profile or any profiles later built on top of them (see the subsequent sections) is therefore also capable of decoding Constrained Baseline profile bitstreams.

3.6.2 Fidelity Range Extensions (Frext) (4 Profiles)

While having a broad range of applications, the first version of AVC was primarily focused on consumer-oriented entertainment-quality video, based on 8-bits/sample and 4:2:0 chroma sampling. For applications such as content-contribution, content-distribution, studio editing and post-processing, it may be necessary to also provide some of the following:

- More than 8 bits per sample of source video accuracy
- Higher resolution for color representation than what is typical in consumer applications – i.e., to use 4:2:2 or 4:4:4 sampling as opposed to 4:2:0 chroma sampling format
- Source editing function features such as alpha blending
- Very high bit rates and very high resolution
- Very high fidelity – even representing some parts of the video losslessly
- Reversible (lossless) color-space transformations
- RGB color representation

The amendment on these "fidelity range extensions" (FRExt) was finalized in 2004. It also includes the following coding tools which are advantageous for such application domains:

- 8×8 intra prediction modes with spatially-filtered predictors
- 8×8 residual transform block size
- Adaptive block-size selection for the residual spatial transform
- Encoder-specified perceptual-based quantization scaling matrices

The FRExt amendment produced a suite of three currently-specified new profiles in a family of profiles collectively called the High family of profiles. One additional profile in the family has been added in 2011.

- *High Profile* (HP), supporting 8-bit video with 4:2:0 sampling, addressing high-end consumer use and other applications using high-resolution video without a need for extended chroma formats or extended sample accuracy.

- *High 10 Profile* (Hi10P), supporting 4:2:0 video with up to 10 bits of representation accuracy per sample.
- *High 4:2:2 Profile* (H422P), supporting up to 4:2:2 chroma sampling and up to 10 bits per sample.
- *Progressive High Profile* (PHP), a subset of the High Profile specified in 2011 to exclude the field coding features (field coding, PAFF and MBAFF) not needed in some applications.

All of these profiles support all features of the prior Main profile, and additionally support the four additional features listed above. The High profile has been incorporated into several important application specifications, particularly including HD content compression in the BD-ROM Video specification of the Blu-ray Disc Association, and the DVB (digital video broadcast) standards for European broadcast television. Indeed, the High profile has rapidly overtaken the Main profile in terms of implementation interest. This is because the High profile adds more coding efficiency to what was previously defined in the Main profile, without adding any significant amount of additional implementation complexity.

3.6.3 Professional Profiles (PP) Extensions (5 Profiles)

After finalization of FRExt, it turned out that the related profiles did not completely fulfill the requirements at the professional quality end of applications. In particular for quality ranges beyond 10 bit/sample and 4:4:4 color sampling, definition of several new profiles became necessary. Further, in this range of qualities, motion-compensated prediction is no longer as important as it is for the lower quality ranges (due to the fact that the less significant bits are partially bearing noise rather than structure). Even more, random frame access is important in editing and contribution, such that the definition of intra-only profiles was reasonable. The new amendment on professional quality extensions which was finalized in 2007 contains five new profiles which can mainly be seen in their relationship with the original "High" family of profiles of FRExt:

- *High 4:4:4 Predictive Profile*, which is a superset of High 4:2:2, supporting up to 14-bit video and 4:4:4 (respectively 4:4:4:4 including alpha channel) color sampling and transform-bypass lossless predictive coding,
- *High 4:4:4 Intra Profile*, which is a subset of High 4:4:4 Predictive, not using motion compensation, and allowing usage of both CAVLC and CABAC entropy coding,
- *CAVLC 4:4:4 Intra Profile*, which is a subset of High 4:4:4 Intra, but not supporting CABAC,
- *High 4:2:2 Intra Profile*, which is a subset of both High 4:4:4 Intra and High 4:2:2 from FRExt, supporting up to 10-bit video, 4:2:2 color sampling and only CABAC entropy coding,

- *High 10 Intra Profile*, which is a subset of both High 4:2:2 Intra and High 10 from FRExt, supporting up to 10-bit video, and 4:2:0 color sampling.

The professional profiles are further discussed in Ref. [9].

3.6.4 Scalable Video Coding (SVC) Extensions (3 Profiles)

Scalable Video Coding (SVC) is an extension of the AVC standard that was finalized in 2007. The concept is to have a video stream which is structured into a base layer stream (being identical to a stream that would conform to a previous AVC profile and as such can be decoded by any conforming decoder), and one or several enhancement layer stream(s) which can only be decoded by a decoder conforming to the respective scalable profile. Separation of the various stream layers is possible from the NAL unit header information without any further decoding. SVC is implementing the following profiles:

- *Scalable Baseline Profile* builds on top of a base layer which is the Constrained Baseline Profile mentioned above; the SVC enhancement layer supports I, P, and B slices (supporting temporal scalability from the key pictures of the base layer), entropy coding CAVLC and CABAC (starting from level 2.1), spatial scalability factors 1:1, 1:1.5 and 1:2,
- *Scalable High Profile* builds on top of a High Profile base layer; the SVC enhancement layer supports I, P, and B slices, interlaced coding tools, entropy coding CAVLC and CABAC, and arbitrary spatial scalability factors; it is also a superset of Scalable Baseline,
- *Scalable High Intra Profile* builds on top of a High Profile base layer that is constrained to using only I slices; the SVC enhancement layer supports I slices, entropy coding using CAVLC and CABAC, and arbitrary spatial scalability factors; it is also a subset of the Scalable High Profile, intended mainly for the area of professional applications.

The SVC extensions are further discussed in Ref. [7].

3.6.5 Multi-view Video Coding (MVC) and 3D Stereo Extensions (2 Profiles)

Two Profiles are defined in the context of MVC:

- *Multiview High Profile* supports only frame coding tools, but with an arbitrary number of views as supported by the given level constraints.
- *Stereo High Profile* supports both frame and field coding tools, but is restricted to two views. The Stereo High Profile was defined about a year after the Multiview High Profile, in response to industry requests to provide such a stereo-only profile with PAFF and MBAFF features supported.

In terms of level limits, MVC repurposes the definitions of single-view AVC decoders for decoding stereo/multiview video bitstreams, such that it is possible to trade off spatio-temporal resolution with number of views. There are some additional constraints to enable multiple parallel decoder implementations of MVC. These extensions and other types of 3D stereoscopic video support using AVC are further discussed in Ref. [8].

3.7 Performance and Applications

Figure 3.14 shows plots of the peak signal to noise ratio (PSNR), which is an often-used measure for objective video quality measurements, comparing AVC against some other designs (using the respective reference software packages developed in the standardization bodies). Compared to MPEG-2, the results indicate that same PSNR is achieved at points of half rate or even less for AVC. This is typical, although it will vary depending on the characteristics of individual video sequences.

However, PSNR figures do not fully reflect visual quality. To provide subjective quality results, a formal subjective test was performed [10] after the finalization of the first version of AVC. Some results for the same sequences from Fig. 3.14 are shown in Fig. 3.15, comparing AVC against the MPEG-4 Part 2 video coding standard (the Advanced Simple Profile using reference software). Results are shown on a five-grade mean opinion score (MOS) scale, based on a variant of a double stimulus continuous quality scale test as described in Ref. [11]. The results indicate that AVC achieves similar or in some cases even better quality as MPEG-4 Part 2 when operating at half of the encoded bit rate.

The following are some particularly important applications of AVC:

- **Mobile telephony**: AVC is shipped with the vast majority of mobile phones that support video.
- **Mobile TV**: AVC is approximately the only video coding standard used in this application e.g. in Finland, Italy, Japan, Korea, Malaysia, Qatar, USA, and many more countries.

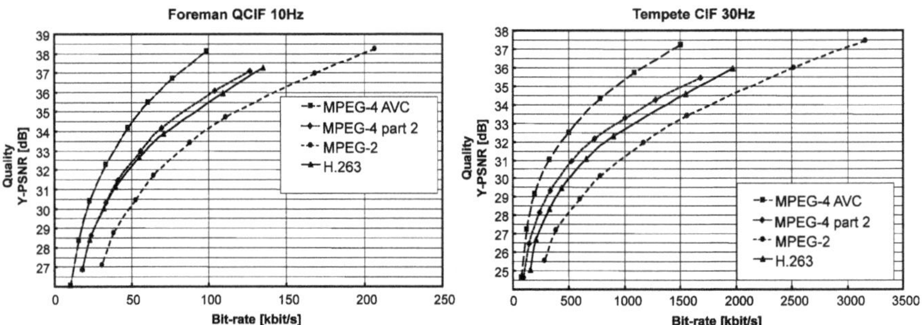

Fig. 3.14 Peak signal-to-noise ratio for the cases of the video sequences Foreman QCIF (*left*) and Tempete CIF (*right*), AVC compared against other standard coding standards

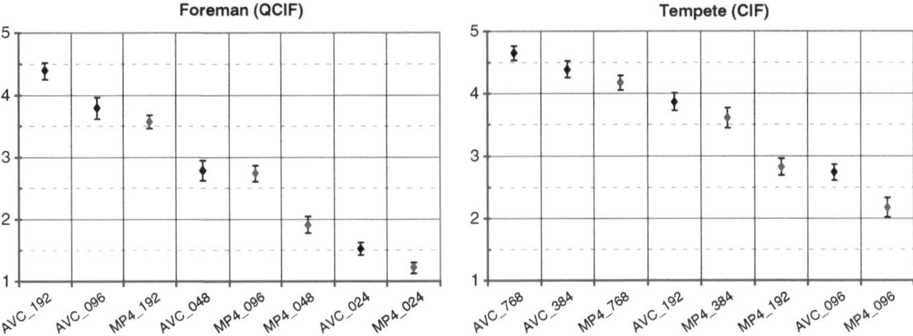

Fig. 3.15 Peak signal-to-noise ratio for the cases of the video sequences Foreman QCIF (*left*) and Tempete CIF (*right*), AVC compared against MPEG-4 Part 2 coding standard (from Ref. [10])

- **Mobile video players**: Apple's iPod and Sony's PlayStation Portable are running AVC on millions of devices.
- **TV broadcast**: **AVC** is used in essentially all recent launches of digital TV broadcast, e.g. in Brazil, Estonia, France, Norway, etc. Practically any HDTV broadcast in the U.S. and Europe is based on AVC, except for over-the-air free broadasts which still use MPEG-2 for legacy compatibility reasons. As an example, DirecTV in the U.S. is broadcasting more than 1,000 HDTV channels – all coded with AVC (and their main competitor, Dish Network, uses it as well). All HDTV services in Germany are using AVC, as well as most other European countries. Worldwide, almost all IPTV services are also based on AVC.
- **Camcorders**: AVC is widely used in digital HD camcorders (often as "AVCHD").
- **Blu-ray disc**: AVC is a mandatory feature in Blu-ray players, and most movies on Blu-ray discs are coded by AVC. Stereoscopic Blu-ray discs use the Stereo High Profile of AVC's MVC extensions.
- **Videotelephony/videoconferencing**: Nearly all current videotelephony and videoconferencing (and "telepresence") systems use AVC. The videoconferencing market was one of the earliest adopters of the standard.
- **Internet video**: Apple's Quicktime, Adobe's Flash, and Google's YouTube, as well as Microsoft's Silverlight, Internet Explorer, Windows 7, Expression Encoder and Media Player all support AVC. It has recently been estimated that the vast majority of video sent over the Internet is encoded using AVC – and that the majority of Internet traffic is now video.

Overall, AVC is the dominant video coding design of today, and it is used in essentially all applications that use video. The core design of the standard has given it exceptionally strong compression capability relative to prior designs together with the flexibility and robustness to enable its use in an extremely broad variety of network and application environments. Its several generations of extensions that have been added to its design have further broadened its capabilities to apply in professional-domain environments and have given it highly-flexible scalability features and 3D stereoscopic and multiview support – while retaining a consistency of design approach that make it straightforward to deploy AVC products that support a broad range of applications. Of course MPEG does not rest on its laurels (and neither does MPEG's collaborative video partner VCEG in the ITU-T), but that is a subject for another chapter.

References

1. ITU-T and ISO/IEC, ITU-T Rec. H.264 | ISO/IEC 14496-10 *Advanced Video Coding* (AVC), May 2003 (with subsequent editions and extensions).
2. J. Ribas-Corbera, P. A. Chou, and S. Regunathan, "A generalized hypothetical reference decoder for H.264/AVC," *IEEE Trans. Circuits Syst. Video Technol.*, vol. 13, pp. 674–687, July 2003.
3. S. Wenger, "H.264/AVC over IP," *IEEE Trans. Circuits Syst. Video Technol.*, vol. 13, pp. 645–656, July 2003.
4. T. Stockhammer, M. M. Hannuksela, and T. Wiegand, "H.264/AVC in wireless environments," *IEEE Trans. Circuits Syst. Video Technol.*, vol. 13, pp. 657–673, July 2003.
5. P. List, A. Joch, J. Lainema, G. Bjøntegaard, and M. Karczewicz, "Adaptive deblocking filter," *IEEE Trans. Circuits Syst. Video Technol.*, vol. 13, pp. 614–619, July 2003.
6. S. Wenger, Y.-K.Wang, T. Schierl, A. Eleftheriadis, "RTP payload format for scalable video coding", IETF document draft-ietf-avt-rtp-svc-27, Feb. 2011.
7. H. Schwarz, D. Marpe, and T. Wiegand, "Overview of the Scalable Video Coding Extension of the H.264/AVC Standard", *IEEE Trans. Circuits Syst. Video Technol.,* vol. 17, no. 9, pp. 1103–1120, Sept. 2007.
8. A. Vetro, T. Wiegand, and G. J. Sullivan, "Overview of the Stereo and Multiview Video Coding Extensions of the H.264/MPEG-4 AVC Standard", *Proc. IEEE*, vol. 99, no. 4, pp. 626–642, Apr. 2011.
9. G. J. Sullivan, H. Yu, S. Sekiguchi, H. Sun, T. Wedi, S. Wittmann, Y.-L. Lee, A. Segall, and T. Suzuki, "New Standardized Extensions of MPEG4-AVC/H.264 for Professional-Quality Video Applications", *Proc, IEEE Intl. Conf. Image Proc.* (ICIP), AntonioSan , TX, vol. I, pp. 13–16, Sept. 2007.
10. ISO/IEC JCT1/SC29/WG11 (MPEG), "Report of The Formal Verification Tests on AVC (ISO/IEC 14496-10 | ITU-T Rec. H.264)", WG11 doc. no. N6231, Waikoloa, HI, Dec. 2003.
11. ITU-R Rec. BT.500-11, *Methods for the Assessment of the Quality of Television Pictures*, June 2002.

Chapter 4
MPEG Video Compression Future

Jörn Ostermann and Masayuki Tanimoto

4.1 Introduction

Looking into the future, more and more of regular and 3D video material will be distributed with increased resolution and quality demand. MPEG foresees further proliferation of high definition video content with resolutions beyond today's HDTV resolutions of 1980×1080 pel. While storage of such video content on solid-state discs or hard discs will not pose a very challenging problem in the future, the distribution of these signals over the Internet, Blu-Ray discs or broadcast channels will, since the expansion of the infrastructure is always an expensive and slow process.

Furthermore, the natural extension of 3D movies is Free Viewpoint Movies where the view changes depending on the position of the viewer and his head orientation.

Based on these predictions, MPEG started two new standardization projects: High Efficiency Video Coding (HEVC) is targeted at increased compression efficiency compared to AVC, with a focus on video sequences with resolutions of HDTV and beyond. In addition to broadcasting applications, HEVC will also cater towards the mobile market.

The second new project 3D video (3DV) supports new types of audio-visual systems that allow users to view videos of the real 3D space from different user viewpoints. In an advanced application of 3DV, denoted as Free-viewpoint Television (FTV), a user can set the viewpoint to an almost arbitrary location and direction, which can be static, change abruptly, or vary continuously, within the limits that are given by the available camera setup. Similarly, the audio listening point is changed accordingly.

J. Ostermann (✉)
Institut fuer Informationsverarbeitung, Leibniz Universität Hannover,
Appelstr. 9A, 30167 Hannover, Germany
e-mail: ostermann@tnt.uni-hannover.de

L. Chiariglione (ed.), *The MPEG Representation of Digital Media*,
DOI 10.1007/978-1-4419-6184-6_4, © Springer Science+Business Media, LLC 2012

4.2 HEVC (High Efficiency Video Coding)

Technology evolution will soon make it possible to capture and display video material with a quantum leap in quality in economic fashion. Here quality is measured in temporal and spatial resolution, color fidelity, and amplitude resolution. Modern TV sets postprocess incoming video to display it at a rate of at least 100 Hz. Camera and display manufactures are showing devices with a spatial resolutions of 4,000 pels/line with 2,000 lines. Each pel can record or display 1024 brightness levels compared to 256 brightness levels today. Use of modern displays enables the display of a wider color gamut than what is used today (Fig. 4.1).

It is difficult in today's transmission networks to carry HDTV resolution with data rates appropriate for high quality to the end user. These higher quality videos will put additional pressure on networks. Future wireless networks like LTE or 4G promise higher bandwidth. However, this bandwidth needs to be shared by a larger number of users making more and more use of their video capabilities. Hence a new video coding standard is required that outperforms AVC at least by 50% and is more suitable for transport over the Internet.

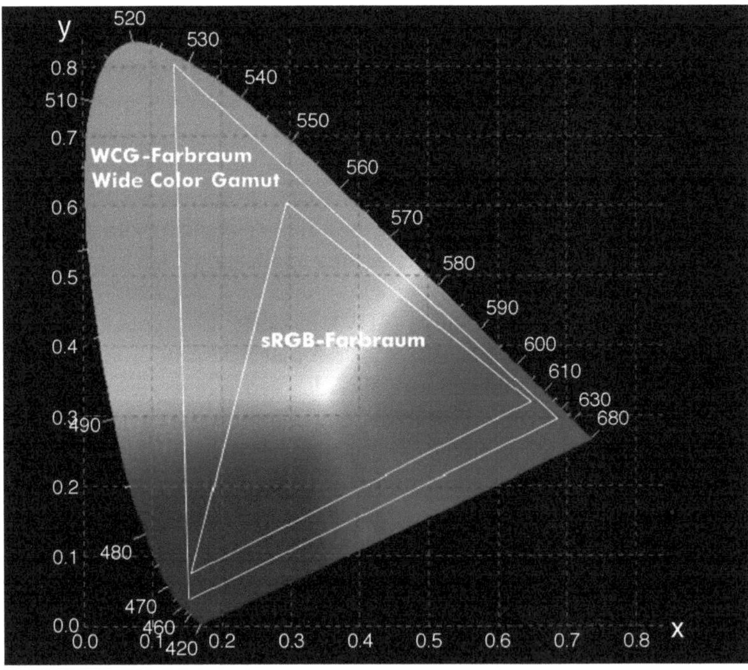

Fig. 4.1 *The shaded area* marks the visible colors, *the triangle sRGB* marks the colors that can typically be displayed on a TV monitor. The *larger Wide Color Gamut triangle* shows the color space of future displays that will be able to display deeper, more saturated *yellows* and *greens*

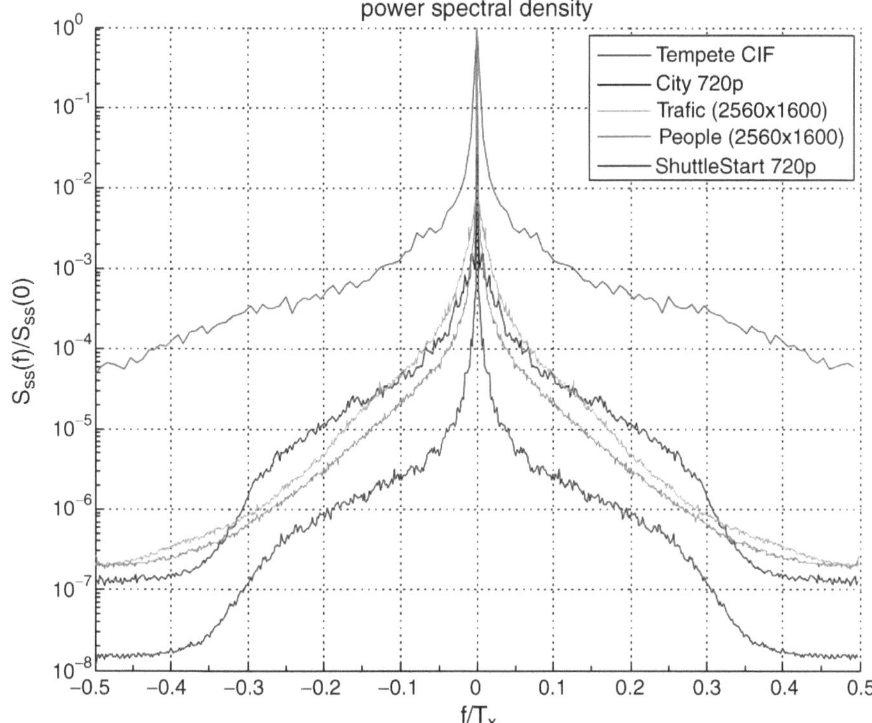

Fig. 4.2 Power spectral density of video sequences with different spatial resolutions showing that high resolution cameras produce less energy at high frequencies compared to low resolution cameras. The legend is valid at f/T = 0.2 from top to bottom

The goal of a 50% gain in coding efficiency will be made possible due to modern video cameras that have different statistical properties compared to cameras produced in the last millennium (Fig. 4.2).

The HEVC video compression standard is currently under joint development by the ISO/IEC Moving Picture Experts Group (MPEG) and ITU-T Video Coding Experts Group (VCEG). MPEG and VCEG have established a Joint Collaborative Team on Video Coding (JCT) to develop the proposed HEVC. Sometimes, this group is referred to as JCT-VC.

4.2.1 Application Scenarios

MPEG envisions HEVC to be potentially used in the following applications: Home and public cinema, surveillance, broadcast, real-time communications including video chat and video conferencing, mobile streaming, personal and professional storage, video on demand, Internet streaming and progressive download, 3D video,

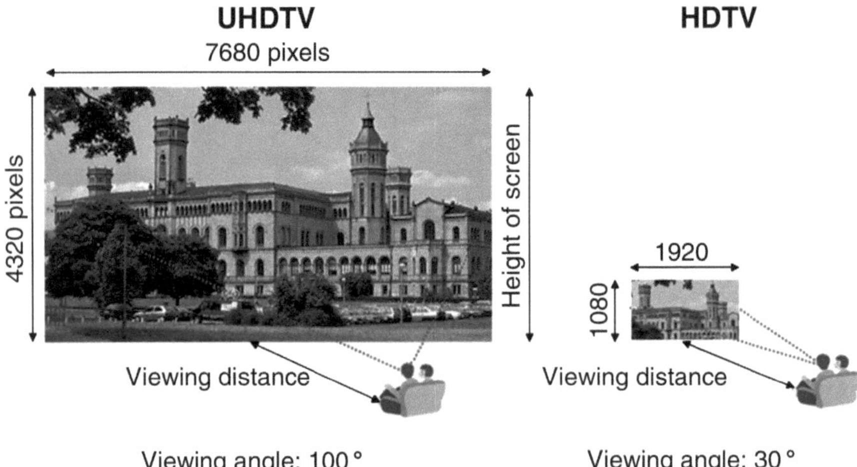

Fig. 4.3 Home theater: Assuming a screen height of 1 m, the viewing distance is 3 m for HDTV and 0.75 m for UHDTV

content production and distribution as well as medical imaging. Looking at this list of applications, the differentiation to AVC and MPEG-2 will be the higher quality of the recorded and delivered video at lower bitrates as well as the better performing streaming services for the Internet enabling real-time communications, video on demand, and Internet streaming. Given these performance improvements, the following applications will be the main applications driving the use of HEVC:

- Broadcast of video services is constantly suffering from bandwidth limitations. The number of programs delivered over the air is severely restricted. Due to the limited bandwidth, HDTV broadcast is not available in many markets. Introduction of HEVC will enable broadcast over the air in these markets. Satellite and cable will follow such that customers can make the most out of their ultra-high definition displays.
- Home theater is a dream of many home owners. New residential buildings often have a room for home theater which will enable the new screen sizes and viewing distances possible with ultra high definition TV (Fig. 4.3). The owners of these rooms tend to spend money on buying the latest and best devices and contents.
- IPTV of video services today requires special networks where only the owner of the network is able to provide IPTV services or IPTV services are offered at lower quality by service providers that do not own the network. Verizon and German Telekom are network owners offering HDTV IPTV at high quality, Netflix as an example for a content owner delivers HDTV at less than 4Mbit/s resulting in limited quality. Reducing the data rate of coded content or increasing quality at today's bitrates will create another competitive market for delivery of TV and Video on Demand services.

Terrestrial broadcast of HDTV, delivery of UHDTV as well as IPTV will be the driving force for pushing HEVC into the market. The consumer strives for the best equipment and content quality. The network owners are short of capital to increase the available speed of the network. This is the ideal environment for a new video coding standard to prosper.

4.2.2 Requirements

The requirements that the new standard will fulfill are various. In the following we focus on those metrics that go beyond AVC.

- Compression performance: HEVC will enable a substantially greater bitrate reduction over AVC High Profile. Past experience shows that the success of a new coding standard depends on a substantial differentiation from alternative standards. Therefore, HEVC will have to outperform AVC by 50%, i.e. the same quality will be delivered using half the bitrate.
- Picture formats: HEVC shall support rectangular progressively scanned picture formats of arbitrary size ranging at least from QVGA to 8000×4000 pel. In terms of color, popular color spaces like YCbCr and RGB as well as a wide color gamut will be supported. The bit depth will be limited to 14 bits/component.

  The support for interlaced material is not foreseen. While interlace was important in the past, modern screens always convert interlaced material into progressive picture formats. The artifacts of this conversion as well as the compute power can be avoided when using progressively scanned material.
- Complexity: There are no measurable requirements on complexity. Obviously, the standard has to be implementable at an attractive cost in order to be successful in the market.
- Video bit stream segmentation and packetization methods for the target networks will be developed allowing for efficient use of relevant error resilience measures for networks requiring error recovery, e.g. networks subject to burst errors.

At the end of the standards development process, MPEG will perform verification tests in order to evaluate the performance of HEVC.

4.2.3 Evaluation of Technologies

At the start of the HEVC development process, MPEG and ITU issued a Call for Proposals which invited interested parties to demonstrate the performance of their video codecs on a predefined set of test sequences and bitrates between 256 kbit/s and 14 Mbit/s. The progressively scanned test sequences were recorded using modern video cameras at resolutions including 416×240 pels, 1920×1080 pels, and 4096×2048 pels. Twenty-seven proposals were evaluated by subjective tests. It turned out that for all test sequences at least one codec provided a rate reduction of 50% compared to AVC High Profile. Therefore, JCT-VC is confident that the rate reduction goal will be reached in the time frame of the standards development. The current plan foresees the final approval of the standard by January 2013.

All 27 proposals were based on block-based hybrid coding with motion compensation. Wavelet technology was not proposed. Based on the first evaluation of the available technologies, technologies likely to be part of the new standard were identified. To a large extend, the technologies were components of the five best performing proposals. They were evaluated in an experimental software Test Model Under Consideration (TMUC) until October 2010. In October 2010, the relevant technologies of TMUC were consolidated into TM-H1, which became the common software that is used as the reference for core experiments in the further development of the HEVC standard. TM-H performs about 40% better than the AVC High Profile.

HEVC will provide more flexibility in terms of larger block sizes, more efficient motion compensation and motion vector prediction as well as more efficient entropy coding. To that extend, HEVC will be a further evolutionary step that started with the standard H.261 issued in 1990.

4.3 3DV (3D Video)

A new 3D Video (3DV) initiative is underway in MPEG. 3DV is a standard that targets serving a variety of 3D displays. 3DV develops a new 3DV format that goes beyond the capabilities of existing standards to enable both advanced stereoscopic display processing and improved support for auto-stereoscopic multiview displays.

Here, the meanings of stereo, multiview and free-viewpoint used in 3DV are clarified. Stereo and multiview are words related to the number of captured and displayed views. Stereo means two views and multiview means two or more views. On the other hand, free-viewpoint is a word related to the position of displayed views. Free-viewpoint means the position of displayed views can be changed arbitrarily by users. This is the feature of FTV. View synthesis is needed to realize the free-viewpoint.

Figure 4.4 shows an example of a 3DV system. In Fig. 4.4, the captured views are stereo and the displayed views are multiview. View synthesis is used to generate multiple views at the receiver side, since the number of required views to be displayed is more than the transmitted captured views.

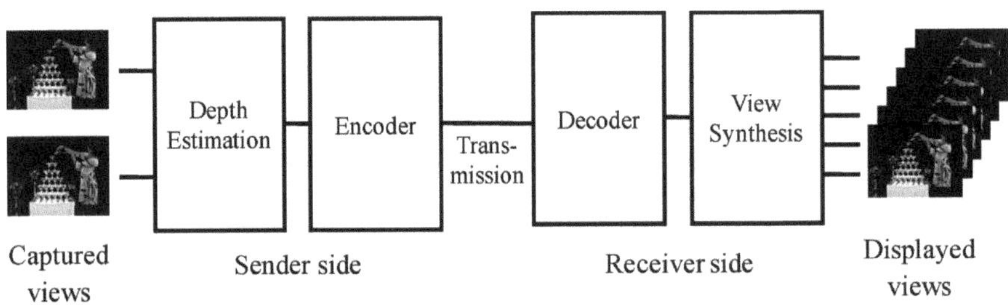

Fig. 4.4 An example of a 3DV system

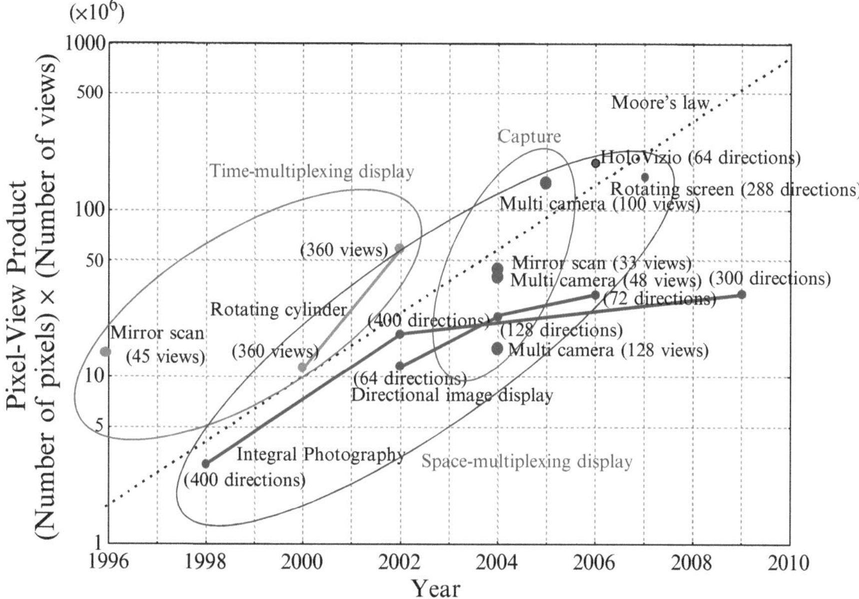

Fig. 4.5 Progress of 3D capture and display capabilities

4.3.1 Background and Motivation

Figure 4.5 shows the progress of 3D capture and display capabilities. In this figure, the ability of 3D capture and display is expressed as a factor of the pixel-view product, defined as "number of pixels" times "number of views". It is seen that the pixel-view product has been increasing rapidly year after year in both capture and display. This rapid progress indicates that not only two-view stereoscopic 3D but also advanced multi-view 3D technologies are maturing.

Taking into account such development of 3D technologies, MPEG has been conducting 3D standardization activities as shown in Fig. 4.6. MPEG-2 MVP (Multi-View Profile) was standardized to transmit two video signals for stereoscopic TV in November 1996. After intensive study on 3DAV (3D Audio Visual), the standardization of MVC that enables efficient coding of multi-view video started in March 2007. It was completed in May 2009. MVC was the first phase of FTV (Free-viewpoint Television). Before completing MVC, 3DV started in April 2007. It uses the view generation function of FTV for 3D display applications. 3DV is the second phase of FTV. The primary goals are the high-quality reconstruction of an arbitrary number of views for advanced stereoscopic processing functionality and to support auto-stereoscopic displays.

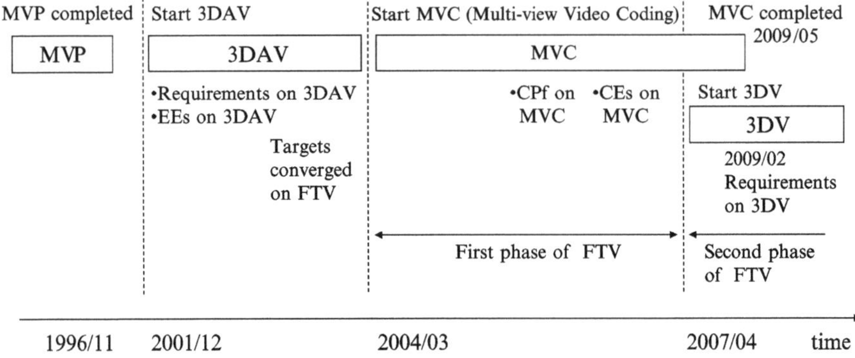

Fig. 4.6 3D standardization activities in MPEG

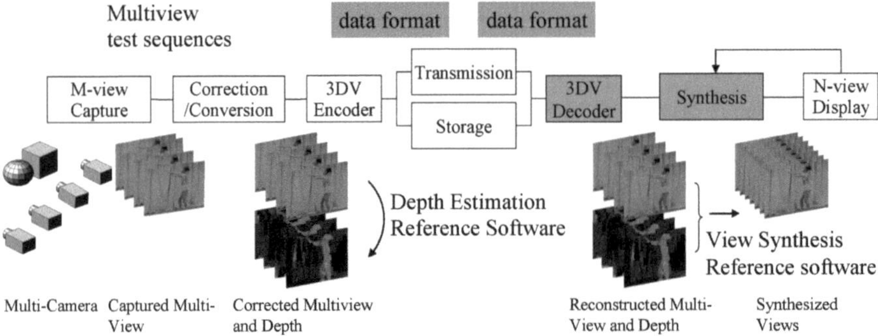

Fig. 4.7 3DV reference model with items considered for standardization

4.3.2 Application Scenarios

The 3DV targets two specific application scenarios.

1. Enabling stereo devices to cope with varying display types and sizes, and different viewing preferences. This includes the ability to vary the baseline distance for stereo video to adjust the depth perception, which could help to avoid fatigue and other viewing discomforts.
2. Support for high-quality auto-stereoscopic displays, such that the new format enables the generation of many high-quality views from a limited amount of input data, e.g. stereo and depth.

4.3.3 Requirements

The 3DV reference model is shown in Fig. 4.7. The input is M views captured by cameras, and the output is N views to be displayed. N can be different from M.

At the sender side, a 3D scene is captured by M multiple cameras. The captured views contain the misalignment and luminance differences of the cameras. They are corrected, and depth for each view is estimated from the corrected views. The 3DV encoder compresses both the corrected multiview and depth, for transmission and storage.

At the receiver side, the 3DV decoder reconstructs the multiview and depth. Then, N views are synthesized from the reconstructed M views with the help of the depth information, and displayed on an N-view 3D display.

Multiview test sequences, depth estimation reference software, and view synthesis reference software are developed in the 3DV standardization activity. They are described in Sect. 4.3.4. Candidate items for standardization are illustrated as blue boxes. Major requirements for each item are shown below.

4.3.3.1 Requirements for Data Format

1. *Video data*
   The uncompressed data format shall support stereo video, including samples from left and right views as input and output. The source video data should be rectified to avoid misalignment of camera geometry and colors. Other input and output configurations beyond stereo should also be supported.
2. *Supplementary data*
   Supplementary data shall be supported in the data format to facilitate high-quality intermediate view generation. Examples of supplementary data include depth maps, segmentation information, transparency or specular reflection, occlusion data, etc. Supplementary data can be obtained by any means from a predetermined set of input videos.
3. *Metadata*
   Metadata shall be supported in the data format. Examples of metadata include extrinsic and intrinsic camera parameters, scene data, such as near and far plane, and others.

4.3.3.2 Requirements for Compression

1. *Compression efficiency*
   Video and supplementary data should not exceed twice the bit rate of state-of-the-art compressed single video. It should also be more efficient than state-of-the-art coding of multiple views with comparable level of rendering capability and quality.
2. *Synthesis accuracy*
   The impact of compressing the data format should introduce minimal visual distortion on the visual quality of synthesized views. The compression shall support mechanisms to control overall bitrate with proportional changes in synthesis accuracy.

3. *Backward compatibility*
   The compressed data format shall include a mode which is backwards compatible with existing MPEG coding standards that support stereo and mono video. In particular, it should be backwards compatible with MVC.
4. *Stereo/mono compatibility*
   The compressed data format shall enable the simple extraction of bit streams for stereo and mono output, and support high-fidelity reconstruction of samples from the left and right views of the stereo video.

4.3.3.3 Requirements for Rendering

1. *Rendering capability*
   The data format should support improved rendering capability and quality compared to existing state-of-the-art representations. The rendering range should be adjustable.
2. *Low complexity*
   The data format shall allow real-time synthesis of views.
3. *Display types*
   The data format shall be display-independent. Various types and sizes of displays, e.g. stereo and auto-stereoscopic N-view displays of different sizes with different number of views shall be supported.
4. *Variable baseline*
   The data format shall support rendering of stereo views with a variable baseline.
5. *Depth range*
   The data format should support an appropriate depth range.
6. *Adjustable depth location*
   The data format should support display-specific shift of depth location, i.e., whether the perceived 3D scene (or parts of it) are behind or in front of the screen.

4.3.4 Available Technologies

4.3.4.1 Multiview Test Sequences

Excellent sets of multiview test sequences are available. Several organizations captured various indoor and outdoor scenes with stationary and moving multiview cameras. The multiview cameras are placed on a straight line and face front in parallel. This camera setting is denoted by 1D parallel in the following. The misalignment and color difference of the cameras are corrected. The corrected multiview test sequences with avail-able depth map data are listed below. Contact each organization and follow the conditions to use them.

1. Nagoya University Data Set (three indoor, two moving camera)
   Pantomime (indoor, 80 views, large depth range, colorful), Champagne_tower (indoor, 80 views, reflections, thin objects, transparency), Dog (in-door, 80 views),

Kendo (moving camera, seven views, colorful, fast object motion, camera motion), Balloons (moving camera, seven views, fast object motion, camera motion, smoke)
2. HHI Data Set (three indoor, one outdoor)
   Book_arrival (indoor, 16 views, textured background, moving narrow objects), Leaving_laptop (indoor, 16 views, textured background, moving narrow objects), Doorflowers (indoor, 16 views, textured background, moving narrow objects), Alt-Moabit (outdoor, 16 views, traffic scene)
3. Poznan University of Technology Data Set (two moving camera, two outdoor)
   Poznan_Hall1 (moving camera, nine views, large depth range, camera motion), Poznan_Hall2 (moving camera, nine views, large depth range, camera motion, thin objects), Poznan_Street (outdoor, nine views, traffic scene, large depth range, reflections and transparency), Poznan_CarPark (outdoor, nine views, large depth range, reflections and transparency)
4. GIST Data Set (two indoor)
   Newspaper (indoor, nine views, rich in texture, large depth range), Cafe (indoor, five views, rich in texture, large depth range, low-res depth captured by five depth-cameras)
5. ETRI/MPEG Korea Forum Data Set (two outdoor)
   Lovebird1 (outdoor, 12 views, colorful, large depth range), Lovebird2 (outdoor, 12 views, colorful, large depth range)
6. Philips Data Set (one CG, one indoor)
   Mobile (CG, five views, combination of a moving computer-graphics object with captured images, ground truth depth), Beer Garden (indoor, two views, colorful, depth obtained through stereo-matching combined with blue-screen technology)

4.3.4.2 Depth Estimation Reference Software

The Depth Estimation Reference Software (DERS) has been developed collaboratively by experts participating in the activity. Although stereo matching is used to estimate depth, two views are not enough to handle occlusion. Therefore, the software uses three camera views to generate a depth map for the center view. DERS requires the intrinsic and extrinsic camera parameters and can support 1D parallel and non-parallel camera setups.

When a 3D scene is captured by multiple parallel cameras, a point in the 3D scene will appear at a different horizontal location in each camera image. This gives horizontal disparity. The depth is inversely proportional to the disparity. The disparity is estimated by determining the correspondence between pixels in the multiple images. The correspondence is expressed by matching cost energy. Generally, this energy consists of a similarity term and a smoothing term. The smoothing term stimulates disparity to change smoothly within objects. The most likely disparity for every pixel can be obtained by minimizing this matching cost energy. DERS uses Graph Cuts as a global optimization method to obtain the global minimum rather than a local minimum. To handle occlusions, the similarity term is calculated by matching between the center and left views, and the center and right views, and then the smallest term is selected.

Temporal regularization is applied to the matching cost energy for static pixels to improve the temporal consistency. Furthermore, the reference software supports segmentation and soft-segmentation based depth estimation.

We have also developed a semi-automatic mode of the depth estimation. In this mode, manually created supplementary data is input to help the automatic depth estimation to obtain more accurate depth and clear object boundaries.

4.3.4.3 View Synthesis Reference Software

The View Synthesis Reference Software (VSRS) has been developed collaboratively by experts participating in the activity.

Since a virtual view between two neighboring camera views is generated, VSRS takes two views, i.e. reference views, two depth-maps, configuration parameters, and camera-parameters as inputs, and synthesizes a virtual view between the reference views. VSRS requires the intrinsic and extrinsic cam-era parameters and can support 1D parallel, and non-parallel camera setups in 1D-mode and General-mode, respectively.

In General-mode, the left and right depth-maps are warped to the virtual view, and both virtual depths are filtered. These depth maps are used to warp the left and right reference views to the virtual view. Holes caused by occlusion in each warped view are filled by pixels from the other view. The warped images are blended and any remaining holes are filled by inpainting.

In 1D-mode the left and right reference views are warped to the virtual view using image shifting. Several modes of view blending and hole filling are supported which consist of different combinations of z-buffering and pixel splatting.

To reduce visible artifacts around object edges, a boundary noise removal method is implemented.

4.4 Summary

With the upcoming standards HEVC and 3DV, MPEG and JCT-VC will provide the codecs to deliver highest quality video content in 2D and 3D. Due to the limitation of bandwidth and stereo TV, markets for the new standards will develop quickly.

Chapter 5
MPEG Image and Video Signature

Miroslaw Z. Bober and Stavros Paschalakis

5.1 Overview of the MPEG-7 Standard

MPEG-7, formally called ISO/IEC 15938 Multimedia Content Description Interface, is an international standard aimed at providing an interoperable solution to the description of various types of multimedia content, irrespective of their representation format. It is quite different from standards such as MPEG-1, MPEG-2 and MPEG-4, which aim to represent the content itself, since MPEG-7 aims to represent information about the content, known as metadata, so as to allow users to search, identify, navigate and browse audio–visual content more effectively. The MPEG-7 standard provides not only elementary visual and audio descriptors, but also multimedia description schemes, which combine the elementary audio and visual descriptors, a description definition language (DDL), and a binary compression scheme for the efficient compression and transportation of MPEG-7 metadata. In addition, reference software and conformance testing information are also part of the MPEG-7 standard and provide valuable tools to the development of standard-compliant systems.

The MPEG-7 visual descriptor set [1–3] comprises a number of non-textual descriptors, typically describing statistical features of the underlying visual data. These features may relate to color (e.g. Dominant Color Descriptor and Scalable Color Descriptor), texture (e.g. Homogeneous Texture Descriptor and Edge Histogram Descriptor), shape (e.g. Region-Based Shape Descriptor and Contour-Shape Descriptor) and motion (e.g. Motion Activity Descriptor and Motion Trajectory Descriptor).

M.Z. Bober (✉)
Visual Atoms Ltd., The Surrey Technology Center, 40 Occam Road,
Guildford, Surrey, GU2 7YG, UK

Center for Vision, Speech and Signal Processing,
University of Surrey, Guildford, Surrey, GU2 7XH, UK
e-mail: m.bober@surrey.ac.uk

L. Chiariglione (ed.), *The MPEG Representation of Digital Media*,
DOI 10.1007/978-1-4419-6184-6_5, © Springer Science+Business Media, LLC 2012

The aforementioned MPEG-7 visual descriptors were designed to identify and provide access to similar content, i.e. similar in terms of color characteristics, texture characteristics, etc. As such, they are not well suited for content identification applications where the aim is to identify the same or derived (modified) content. This led to the development of the MPEG-7 Image and Video Signature Tools, collectively known as the MPEG-7 Visual Signature Tools, which were designed specifically for the task of fast and robust content identification. The MPEG-7 Visual Signature Tools are the latest additions to the MPEG-7 standard and are examined in detail in this chapter.

5.2 The Visual Signatures

In recent years, people have been creating and consuming an ever increasing amount of digital image and video content. A recent survey of prominent web sites shows that Flickr has over 2 billion images, Photobucket has over 4 billion and Facebook has 1.7 billion. At the same time, according to the company's own figures, in 2010 people were watching two billion videos a day on YouTube and uploading hundreds of thousands of videos daily, at a rate of 24 h of content every minute. There are hundreds of billions of images and videos on the Internet, and even users' personal databases can have tens or hundreds of thousands of images and videos. However, there are currently few tools which one can use to efficiently identify an image or video or search for a specific piece of content, possibly in an edited or modified form, either on the Internet or in one's own personal collection. The recently standardised MPEG-7 Image Signature Tools and Video Signature Tools address this problem by providing an interoperable solution for image and video content identification.

Unlike previous MPEG-7 descriptors which were designed to provide access to similar content, the Visual Signatures are content-based descriptors designed with the task of content identification in mind, i.e. designed for the fast and robust identification of the same or derived (modified) visual content in web-scale or personal databases. Descriptors of this type are sometimes also known as robust hashes or fingerprints and have a strong advantage over watermarking techniques because they do not require any modification of the content and therefore can be used readily with all existing content.

The applications for the MPEG-7 Visual Signatures are numerous. Some examples include:

- Rights management and monetization: Detection of possible copyright infringement, brand misuse, or content monetization online (for content owners), or identification of the copyright owner to avoid infringement or ensure correct attribution (for content consumers).
- Media usage monitoring: Tracking and recording statistics such as distribution and frequency of content usage, e.g. an advertising agency may want to verify that its advertisements are placed correctly.

- Web-page linking: Visual content can be used to imply links between web-pages, as is currently done for text, reducing the dependence on text annotations and providing additional clues to the semantic structure.
- Metadata association: The ability to quickly and accurately obtain information, such as title, author etc., is a very useful tool for digital content management.
- Personal collection management: Image and video processing software makes it easy for users to edit their visual content, leading to different versions of the same image or video. The MPEG-7 Visual Signatures provide an ideal tool for personal library management and de-duplication.

The following sections outline the technologies behind the Image and Video Signature Tools, and present the MPEG-7 evaluations procedures and results.

5.3 Image Signature

In this section we describe the MPEG-7 Image Signature as specified in the standard [4]. In order to achieve fast searching and high robustness to loss of content (e.g. cropping) two complementary approaches to represent an image are combined: a global signature where the signature is extracted from the entire image and a local approach where a set of local signatures are extracted at salient points in the image. Both global and local descriptors are based on a multi-resolution trace transform and are explained below.

5.3.1 Global Signature

The core technology for the Image Signature is the trace transform [5] with multi-resolution extensions. In order to obtain a robust descriptor of an image or region, a set of lines is projected over the image or region. The set contains lines at varying angles (Θ) and distance from the origin $[d]$ (Fig. 5.1a) and functionals are applied over these lines producing a trace image (Fig. 5.1b). Next, a second functional, known as the diametrical functional, is applied along columns of the trace image to obtain a 1D function known as the circus function (Fig. 5.1c). In the example of Fig. 5.1 the function in (c) is obtained by summing up the values along the columns of the trace image (b).

But how can we efficiently represent an image with the circus function? One of the most important requirements guiding the design of the Image Signature is high-speed matching. To achieve this we select a binary descriptor that can be matched using a simple hamming distance (i.e. comparing individual bits and counting the number of different bits). Hamming distance computation can be implemented very efficiently in both hardware and software implementations, e.g. using XOR, summations and/or lookup-tables. Furthermore, we can achieve rotational invariance without any additional computations thanks to the useful property of the trace

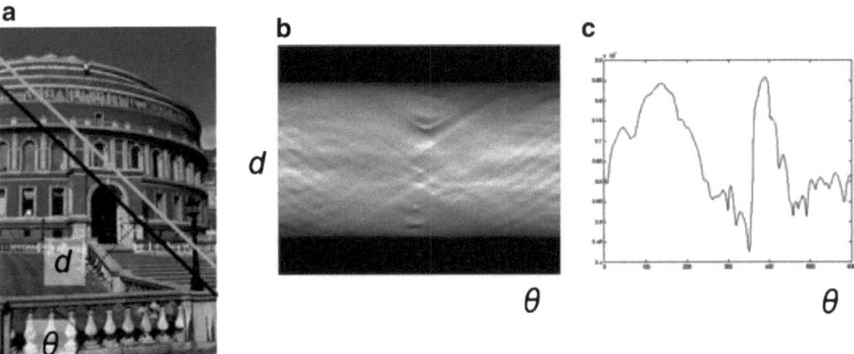

Fig. 5.1 Trace transform of an image, (**a**) original image with lines projected, (**b**) its trace transform, and (**c**) the resulting 1-D circus function

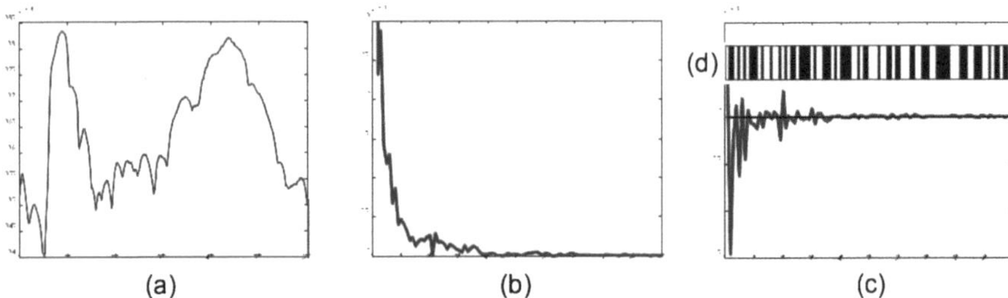

Fig. 5.2 Extraction of a binary descriptor from the circus function, (**a**) circus function, (**b**) magnitude of FFT coefficients of the circus function and the differences between neighbouring coefficients (**c**) with the thresholded binary representation shown as a bar-code above the graph (**d**)

transform which converts rotation of the original image into translation of the circus function. The binary descriptor is obtained from the circus function (see Fig. 5.2) by taking the magnitude of its FFT coefficients (Fig. 5.2b), computing the differences between the neighbouring coefficients and thresholding (Fig. 5.2c). The difference greater then zero corresponds to binary value 1, otherwise binary value 0 is assumed. As a result we obtain a binary string that can be visualised as a barcode (Fig. 5.2d).

In order to obtain a very low false alarm rate, four additional types of information are obtained from the trace transform. These correspond to analysis of relations between variable-width bands, double-cones and circles from the image (see Fig. 5.3). In total 21 different component representations are created including 12 circus functions from multi-resolution representation, five band-circus functions, one trace-circular function and three trace-annulus functions, each producing a 48-bit component binary descriptor.

The extraction process results in 1008 bits (21×48) and a subset of 512 bits is specified by the standard – it was chosen by considering the entropy of the bits on independent data.

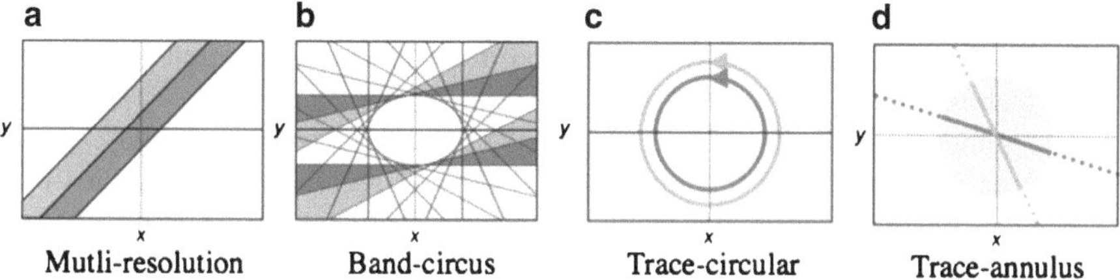

Fig. 5.3 Circus functions are designed to represent relations between image intensities in variable-width bands (**a**), double-cones (**b**), circles (**c**) and lines with limited extend (**d**)

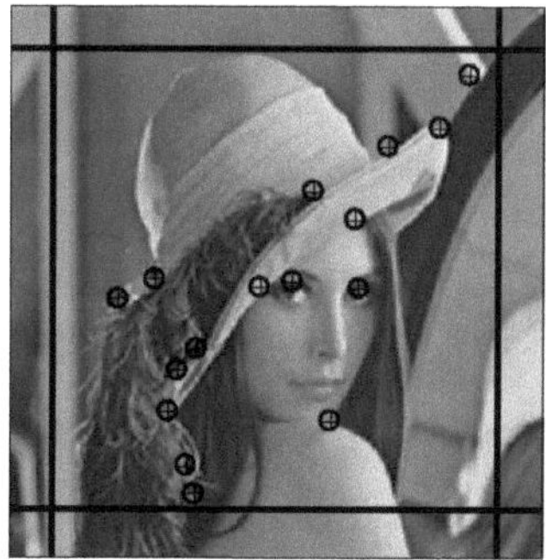

Fig. 5.4 Location of feature points for Lena image

Two global signatures of 512 bits are extracted using the same algorithm but with a different pre-processing (size normalisation) applied to the input image – one on the image where aspect ratio is maintained and one on the image resized to square dimension of 192×192. This second signature provides robustness to aspect ratio changes.

5.3.2 Local Signature

The global signature, while ultra-fast in searching, gives only limited robustness to loss of content (e.g. image cropping up to 20%). To address this limitation, a set of local signatures centered at salient points in the image are used. First and second order image derivatives are analysed to find spatial-scale locations of features.

After aspect-ratio preserving size normalisation, a 12-level scale-space representation of the image is constructed by repeatedly smoothing the pre-processed image with increasing filter size. Features are detected by finding maxima in the location-scale space. Two feature detectors are used to provide good detection of the location and scale of features: (1) the modified scale-corrected Laplacian of Gaussian is used to detect features' locations and scales and (2) the scale-adapted Harris operator is used to reject features with a weaker response. Between 32 and 80 features are selected for each image, excluding locations in areas close to the edge of the image or in the vicinity of already selected stronger features. Locations of feature points for the Lena image are presented in Fig. 5.4. The maximum number of feature points (80) is a trade-of between the fast matching and robustness to cropping. Local descriptors of the image intensity are then extracted from the regions centred at locations of the feature points.

The global trace transform method is applied to extract a binary "feature signature" from regions around feature points. A circular region, having a radius of r (dependent on the scale of the features s) pixels, centered on a feature point is selected and the binary signature is computed. A simplified version of the global signature algorithm is used to extract feature signatures. The location and direction of features are also included as part of the signature to allow for improved independence, robustness and search speeds.

The complete Image Signature is made up of two global signatures and up to 80 local signatures. The global signatures are each 512 bits and the local signatures are each 80 bits giving a complete signature of up to 7424 bits ($2 \times 512 + 80 \times 80$).

5.3.3 Evaluation Process and Performance

For the evaluation process the minimum performance limits were defined for three requirements: (1) independence (false alarm rate), (2) matching speed and (3) descriptor size. The decision process was then based on the matching performance and its robustness to content modifications, expressed as the mean detection rate.

Analysis was carried out on the global image signature and the complete (i.e. local and global) image signatures to find the thresholds (algorithm parameters) corresponding to a false acceptance of 0.05 ppm (ppm) and 10 ppm respectively. A database of over 135,000 images was used to determine the independence thresholds, resulting in over nine billion possible image pairs. The dataset is made up of colour photographs, greyscale photographs, logos and graphics of varying sizes. The robustness for the global and complete image signatures are provided in Tables 5.1 and 5.2. A study was carried out to compare the global signature with several other state-of-the-art methods and is presented in [6]. It shows that the performance of the Image Signature is above that of the alternative approaches.

Search speeds achieved are around 80 million global signatures per second and 100,000 complete signatures per second on a single core of an Intel Xeon 3 GHz processor. This speed performance is measured based on pair-wise matching, without any indexing, as indexing is not possible for some application scenarios. Naturally, for some application where database indexing if feasible, the search speed can be significantly increased. The searching is also scalable so can be implemented on multi-processor environments.

Reference software and a conformance testing set of bitstreams have been developed along with the standard [7, 8]. They can be used to alleviate the development of devices that shall comply with the standard.

Table 5.1 Average success rate for the global image signature

Modification type	Modification level			Mean
	Heavy	Medium	Light	
Brightness change	99.67%	99.80%	99.96%	99.81%
Colour –>monochrome conversion			99.99%	99.99%
JPEG compression	99.68%	99.94%	100%	99.87%
Colour reduction		99.35%	99.84%	99.60%
Gaussian noise	98.90%	99.53%	99.90%	99.77%
Histogram equalisation			93.32%	93.32%
Autolevels			99.90%	99.90%
Blur	100%	100%	100%	100%
Simple rotation	99.99%	99.86%	99.95%	99.93%
Scaling	100%	100%	100%	100%
Flip (left to right)			100%	100%
Mean				99.29%

Table 5.2 Average success rate for the complete image signature

Modification type	Modification level			Mean
	Heavy	Medium	Light	
Rotation	99.99%	100%	100%	100%
Translation	75.64%	91.85%	99.42%	88.97%
Aspect ratio		100%	100%	100%
Crop	54.95%	91.89%	99.48%	82.11%
Skew	94.86%	99.96%	100%	98.27%
Perspective	96.15%	99.08%	99.85%	98.36%
Combined (crop, translate, scale)	98.33%	99.60%	99.85%	98.26%
Mean (including Table 5.1)				98.04%

5.4 Video Signature

The MPEG-7 Video Signature [9] is a compact and unique content-based video identifier that is robust to typical video editing operations such as on-camera capture, recompression, text/logo overlay, etc., but also sufficiently different for every original content to identify it reliably and uniquely.

5.4.1 Video Signature Extraction

The MPEG-7 Video Signature comprises two main parts: (1) fine signatures extracted from the individual frames of the video, and (2) coarse signatures extracted from sets of fine signatures.

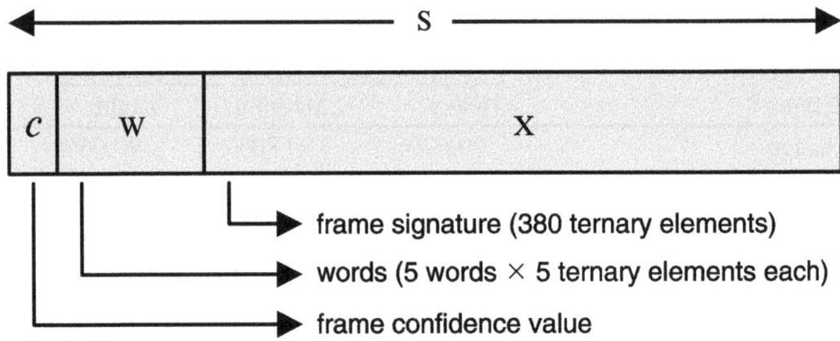

Fig. 5.5 Organization of the frame signature **x**, words **w** and confidence value c in the fine signature **s**

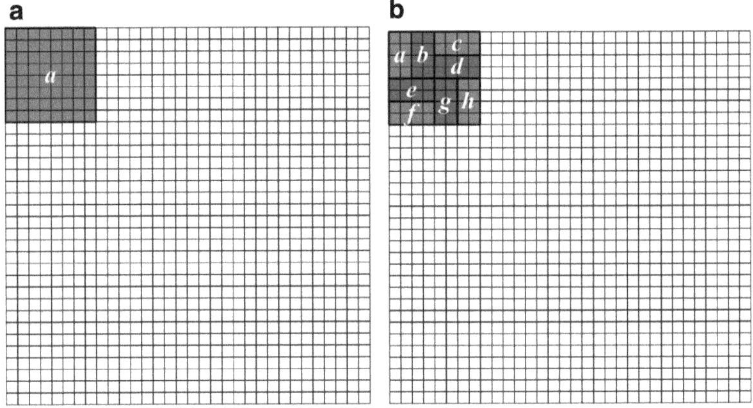

Fig. 5.6 Sample frame signature elements: (**a**) One average element, $\overline{a}$. (**b**) Four difference elements, $\overline{a}-\overline{b}$, $\overline{c}-\overline{d}$, $\overline{e}-\overline{f}$ and $\overline{g}-\overline{h}$

The fine signatures are extracted from 32×32 pixel 8-bit luminance information, derived by the block averaging of the luminance channel of each frame. Each fine signature contains (a) a set of local features, the "frame signature", (b) a small subset of the local features, the "words", and (c) a global "frame confidence" measure. This is illustrated in Fig. 5.5.

The local features of the fine signature are local average intensities and differences at various scales in the frame. Each local feature is called an element or dimension of the frame signature. There are 380 elements in a frame signature, 32 averages and 348 differences. The frame signature elements were designed to capture the local intensity content and intensity interrelations at different regions and scales of a frame. A sample of these elements is shown in Fig. 5.6. Figure 5.6a illustrates a representative average element, calculated as $\overline{a}$, the average luminance of the pixels in region a after down-sampling the frame to 32×32 pixels. Figure 5.6b illustrates four representative difference elements, calculated as $\overline{a}-\overline{b}$, $\overline{c}-\overline{d}$, $\overline{e}-\overline{f}$ and $\overline{g}-\overline{h}$, i.e. as the difference between the average luminance of the corresponding regions. The other frame signature elements are calculated in a similar fashion.

The 380 frame signature elements are divided into different categories. This division is performed according to characteristics such as the element type (average or difference), extraction scale, and pattern type. There are two categories of average elements, A_1 and A_2, with 20 and 12 elements respectively, and eight categories of difference elements, $D_1 \dots D_8$, with 116, 25, 36, 30, 62, 9, 50 and 20 elements respectively. This categorization is significant for the next step of the frame signature extraction, which is the ternarization of the elements.

More specifically, a frame signature comprises ternarized elements, i.e. quantized to three levels. For the average element categories A_1 and A_2, this ternarization proceeds as follows. Let v_i, $i \in \{0,1, \dots, N_j\}$, be an element of category A_j, $j \in \{1,2\}$.

The ternarized element x_i is calculated as

$$x_i = \begin{cases} 2 & \text{if } (v_i - 128) > ThA_j \\ 1 & \text{if } |v_i - 128| \leq ThA_j \\ 0 & \text{if } (v_i - 128) < -ThA_j \end{cases} \tag{5.1}$$

The threshold ThA_j is adaptive and is re-calculated for each frame and for each category A_j as the 33.3% percentile rank of the absolute values $|v_i - 128|$. That is, for each category A_j with N_j elements we calculate $|v_i - 128|$ $\forall i \in \{0,1, \dots, N_j\}$ and sort the results in ascending order. Then, the threshold ThA_j is nth element of the sorted list, where $n = \lfloor 0.333 \cdot N_j \rfloor$.

For the difference element categories $D_1 \dots D_8$ the ternarization proceeds in a similar manner. Let v_i, $i \in \{0,1, \dots, N_j\}$, be an element of category D_j, $j \in \{1, \dots, 8\}$. The ternarized element x_i is calculated as

$$x_i = \begin{cases} 2 & \text{if } (v_i) > ThD_j \\ 1 & \text{if } |v_i| \leq ThD_j \\ 0 & \text{if } (v_i) < -ThD_j \end{cases} \tag{5.2}$$

The threshold ThD_j is again adaptive, re-calculated for each frame and for each category D_j as the 33.3% percentile rank of the absolute values $|v_i|$, as described above. The aim of making the thresholds frame-adaptive and category-adaptive is to achieve a more uniform distribution of the frame signature elements across the three quantization bins. This avoids the situation of a frame signature with diminished information content, e.g. when a video frame is of very poor contrast or too bright. The vector $\mathbf{x}$ of all 380 ternarized elements is the frame signature.

In the context of the Video Signature, a "word" refers to a compact representation of the complete frame signature and is an ordered subset of elements of the vector $\mathbf{x}$, i.e. a projection from a 380-dimensional space to a lower-dimensional space. For two video frames, the distance between two corresponding words is an approximation of the distance between the complete frame signatures. The Video Signature uses Q Ψ-dimensional words with $Q=5$ and $\Psi=5$, i.e. five different projections from the original 380-dimensional space to a 5-dimensional space. The five words **w** provide a good and compact representation of the complete frame signature and are used at a later stage for the formation of the coarse signature.

The global frame confidence value is an 8-bit integer value, calculated by taking the absolute value of all 348 difference elements and sorting them in a single list and in ascending order. The 174th element of the sorted list, denoted by M, is selected as the median value and the global frame signature confidence c is calculated as $c = \min\left(\lfloor M \cdot 8 \rfloor, 255\right)$.

The frame confidence c, the words **w** and the frame signature **x** make up the fine signature **s** of the Video Signature. In the binary representation, the ternary elements of the frame signature and the words are not represented by two bits each but are encoded. More specifically, each group of five consecutive elements is encoded into an 8-bit value, resulting in a 20% size reduction. Therefore, the fine signature described here is very compact at only 656 bits of storage. In practice, as part of the complete Video Signature for a video segment, the fine signatures require significantly less space for storage by exploiting temporal redundancies, as will be seen later.

The coarse signatures are extracted from sets of fine signatures based on a "bag-of-words" approach [10]. This bag-of-words representation is extracted for temporal segments of 90 consecutive frames. As described earlier, every fine signature contains five words, i.e. five subsets of the complete frame signature. For each of these five words, the values that it takes over the 90 frame sequence are plotted into a 243-bin histogram ($3^5=243$). Therefore, five histograms h_k, $k \in \{1, \ldots, 5\}$, are generated, one for each of the five subsets of the frame signature. Then, each histogram is binarized, by setting the value of each bin to 1 if it is greater than 0, and leaving as 0 otherwise. This gives rise to five binary occurrence histograms b_k, $k \in \{1, \ldots, 5\}$, which become the coarse segment signature **b** for the 90-frame segment. In the Video Signature coarse signatures are generated for 90-frame segments with a 45-frame overlap, e.g. if a coarse signature is extracted for frames m to $m+89$, then another coarse signature will be extracted for frames $m+45$ to $m+134$, etc. This is illustrated in Fig. 5.7. Each coarse signature requires 1215 bits for storage which, given the coarse signature overlap, produces in 810 bits/s at 30 frames/s.

5.4.2 Video Signature Compression

The encoding of ternary values into bytes provides a certain amount of reduction to the size of each frame signature independently. However, there is still considerable temporal redundancy in the Video Signature. This is eliminated through a specially designed lossless low-complexity compression scheme based on predictive coding, run-length coding and entropy coding. More specifically, the compression is applied to consecutive 45-frame temporal segments, which are aligned with

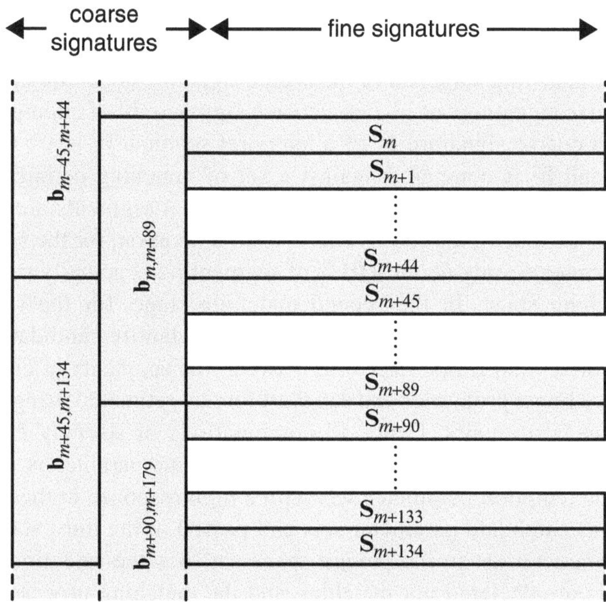

Fig. 5.7 Extraction pattern of coarse signatures **b** from the fine signatures **s**

the coarse segment signature temporal segments. For each temporal segment, only the frame signatures of that segment are compressed, i.e. the compression scheme is not applied to the coarse signatures or the words or frame confidence of the fine signatures. This compression scheme results in a reduction of the size of the frame signature blocks by approximately 75% on average. Besides achieving a high compression performance, the Video Signature compression scheme is also designed to lead to very low complexity implementations and low memory requirements for both encoder and decoder. It should also be noted that, by aligning the compression temporal segments with the coarse segment signature temporal segments, and by leaving the coarse segment signatures uncompressed, only a small fraction of the compressed segments will require decompression during the matching process, which follows a coarse-to-fine approach and is described below. For example, for the MPEG-7 dataset and experiments, only about 0.01% of the compressed segments need to be decompressed.

5.4.3 Video Signature Matching

The standard does not specify the matching method to be used in different applications. This section briefly outlines the matching method that was used during the MPEG-7 evaluation process. The matching between two Video Signatures v^1 and v^2 is carried out in three stages. These stages have been designed to maximize matching speed and true positives and to minimize false positives.

In the first matching stage, all of the coarse signatures of v^1 are compared with all of the coarse signatures of v^2. For a coarse signature $\mathbf{b}^1$ of a temporal segment $f^1 \in v^1$ and a coarse signature $\mathbf{b}^2$ of a temporal segment $f^2 \in v^2$, the similarity between $\mathbf{b}^1$ and $\mathbf{b}^2$ is compared against a set of matching conditions. If these matching conditions are not satisfied, these temporal segments are declared not matching and the matching process ends. As an indication, for the MPEG-7 dataset and experiments, only about 0.01% of segment pairs actually progress to the second matching stage. In the second matching stage, for the segment pairs passed to this stage, a Hough transform is used to identify candidate parameters of temporal offset and frame rate ratio between the segments of each candidate pair. These are linear properties and can therefore be estimated using two strongly corresponding frame pairs. Using all combinations of strongly corresponding frame pairs, a vote is cast to the calculated candidate parameters in the Hough space, and the temporal parameter sets with a high response in the Hough space are selected as candidate parameter sets and passed to the third stage of matching. If no parameter set in the Hough space satisfies the selection criteria, the segment pairs are declared not matching and the matching process ends. In the third stage of matching, the matching interval, i.e. the start and end position of the match between the two segments, is determined by temporal interval growing based on a frame-by-frame matching using the complete frame signatures and the candidate temporal parameter sets identified in the previous matching stage. First, the estimated temporal offset is used to determine the initial temporal matching position. Then, using the estimated frame rate ratio, the temporal interval is extended frame-by-frame towards both temporal directions until the frame signature distance exceeds a threshold. If the length of the resultant matching interval is shorter than a given minimum duration, the matching interval is eliminated as a non-match. Otherwise, the frame confidence element associated with each frame in the matching interval is checked to verify the match. This process is carried out for all of the candidate temporal parameter sets, thus generating multiple candidates of matching intervals. Then, one candidate interval is selected as the best matching result, based on the mean distance of the frame signatures and the interval length.

5.4.4 Evaluation Process and Performance

All technologies proposed to MPEG were evaluated by setting limits for three requirements, namely (1) independence (false alarm rate), (2) matching speed, and (3) descriptor size, and then deciding based on the performance for the robustness requirement, expressed as the mean success rate. For a given modification, the success rate R is defined as $R = C/T$, where C is the number of correct matches found and T is the number of videos that match.

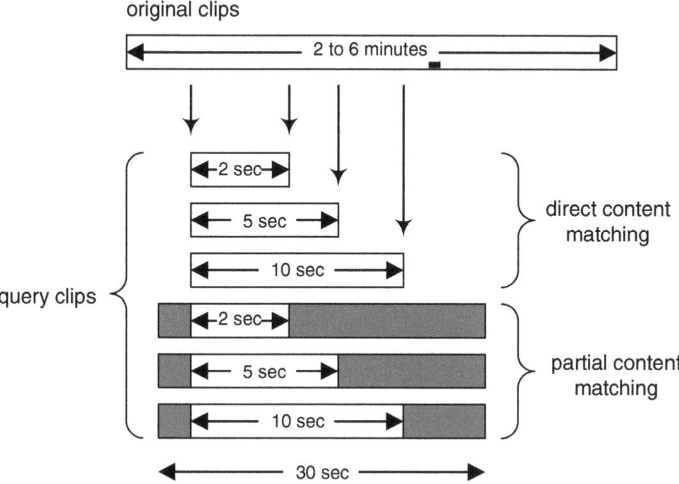

Fig. 5.8 The different query scenarios, referred to as direct or partial matching. Tests are performed at temporal granularities of 2, 5 and 10 s

There are two query types under which the Video Signature Tools were evaluated. The first is direct content matching, where the whole query clip matches with a certain part of the original clip. The second is partial content matching, where only a part of the query clip matches with a certain part of the original clip. Each of the two broad query types has three specific query scenarios to evaluate performance at different temporal granularities. The query scenarios correspond to three durations D of the segment to be matched, i.e. $D=2$, 5 and 10 s. In the case of partial content matching, D indicates the minimum durations of the segment to be matched; the total duration of the query clip is 30 s. The different query scenarios are illustrated in Fig. 5.8.

The requirement for matching speed was that an algorithm will, on average, match at least 1,000 clip pairs per second under the partial content matching scenarios. The requirement for descriptor size was that it shall not exceed 30,720 bits/s of content at 30 frames/s. The requirement for independence was that an algorithm shall achieve a false alarm rate of ≤5 ppm (parts per million), i.e. no more than 5 false matches per million comparisons. In the independence test, each clip in a database of 1,900 3-min clips unrelated to each other was used to produce the required six queries, which were then compared against all the other clips to determine the operational settings for the ≤5 ppm false positive rate. In the robustness test, a database of up to 545 approximately 3-min clips of various types (e.g. film, news, documentary, cartoons, sport, home video, etc.) was used, and each clip generated the six required query clips. Each of these queries was then subjected to one of nine modification categories, with each category having between one and three intensity levels, giving 22 different modifications. The complete list of modifications is shown in Table 5.3. The modified query clips were then compared with the original clips at

Table 5.3 List of modifications in the MPEG-7 robustness tests

	Modification level		
Modification type	Heavy	Medium	Light
Text/logo overlay	30% of screen	20% of screen	10% of screen
Compression at CIF resolution	64 kbps	256 kbps	512 kbps
Resolution reduction from SD	N/A	QCIF	CIF
Frame-rate reduction from 25/30 fps	4 fps	5 fps	15 fps
Camera capture at SD resolution	10% extra border	5% extra border	0% extra border
Analog recording and recapture	3 times	2 times	1 time
Color to monochrome	N/A	N/A	[a]
Brightness change (additive)	+36	−18	+9
Interlaced/progressive conversion	N/A	N/A	P→I→P or I→P

[a] $I = 0.299R + 0.587G + 0.114B$

Table 5.4 Average success rate of the Video Signature in the robustness tests at a false alarm rate of ≤5 ppm

Matching scenario		Average success rate
Direct	2 s	96.65%
	5 s	97.76%
	10 s	98.22%
Partial	2 s	91.73%
	5 s	93.43%
	10 s	95.12%
All		95.49%

the operational parameters determined in the independence test. For the direct content matching, for a match to be correctly detected the detected start point had to be within 1 s of the actual start point. For partial content matching (1) the detected start point in the original clip had to be within 1 s from the true start point, (2) the detected start point in the query clip had to be within 1 s of the actual start point, and (3) the detected duration of the matching part had to be within 2 s of the actual duration of the matching part.

The success rates of the MPEG-7 Video Signature are shown in Table 5.4. The overall success rate of the technology is 95.49%. This is achieved at a false acceptance rate of ≤5 ppm, average matching speed of over 1,500 matches per second, and with an uncompressed descriptor size of approximately 8 MB/h of content, which is reduced to approximately 2 MB/h when using the dedicated descriptor compression scheme.

As for the Image Signature, reference software and conformance testing conditions have also been developed for the Video Signature and are awaiting publication, in order to ease the development of systems that shall comply with the standard.

5.5 Conclusions

The Visual Signature Tools make a valuable addition to the suite of existing MPEG-7 visual descriptors. Unlike the previous MPEG-7 visual descriptors, the Image and Video Signature Tools were designed with the task of content identification in mind, i.e. designed for the fast and robust identification of the same or derived (modified) visual content in web-scale or personal databases. These newly standardised Visual Signatures are the result of extensive collaborative effort within the MPEG-7 group, with the common aim of delivering optimised interoperable content identification solutions, and it is anticipated that they will find wide adoption in such applications.

References

1. ISO/IEC 15938–3 2002, Information Technology – Multimedia content description interface – Part 3: Visual
2. ISO/IEC 15938–3 2002/Amd 1:2004, Information Technology – Multimedia content description interface – Part 3: Visual, Amendment 1: Visual Extensions
3. ISO/IEC 15938–3 2002/Amd 2:2006, Information Technology – Multimedia content description interface – Part 3: Visual, Amendment 2: Perceptual 3D Shape Descriptor
4. ISO/IEC 15938–3:2002/Amd 3:2009, Information Technology – Multimedia content description interface – Part 3: Visual, Amendment 3: Image signature tools
5. A. Kadyrov and M. Petrou, "The Trace Transform and its Applications," *IEEE Transactions on Pattern Analysis and Machine Intelligence*, vol. 23, no. 8, pp. 811–828, August 2001
6. Brasnett, P., Bober, M., "Fast and Robust Image Identification" *In 'Proc. 10th International Conference on Pattern Recognition*, 2008, pp. 1–5
7. ISO/IEC 15938–6 2003/Amd 3:2010, Information Technology – Multimedia content description interface – Part 6: Reference Software, Amendment 3: Reference Software for Image Signature Tools
8. ISO/IEC 15938–7 2003/Amd 5:2010, Information Technology – Multimedia content description interface – Part 6: Conformance Testing, Amendment 5: Conformance Testing for Image Signature Tools
9. ISO/IEC 15938–3:2002/AMD 4:2010, Information Technology – Multimedia content description interface – Part 3: Visual, Amendment 4: Video signature tools
10. D. D. Lewis, "Naive (Bayes) at Forty: The Independence Assumption in Information Retrieval", *In Proc. 10th European Conference on Machine Learning*, Chemnitz, DE, April 21–24 1998, pp. 4–15

Chapter 6
MPEG Audio Compression Basics

Marina Bosi

6.1 Introduction and Background

Audio coding in general and more specifically perceptual audio coding has been one of the most exciting and commercially successful applications of digital signal processing. While started in the mid/late eighties out of sheer necessity for transmission and storage of digital audio and somewhat following in the footsteps of speech compression technologies – one may recall the seminal paper by Schroeder et al. [1] – its commercial success went way beyond anyone's expectations, changing forever the way we experience music. Just looking at the market penetration of DVDs, portable audio devices, internet music applications, and social networking based on musical preferences, and witnessing how music consumption has evolved due to audio coding technology makes one marvel at the impact of this technology and its repercussions on our everyday life.

The first MPEG standard MPEG-1, "Coding of Moving Pictures and Associated Audio for Digital Storage Media at up to about 1.5 Mbit/s" [2], marked the establishment of the first international standard for audio coding. The goal of MPEG-1 Audio was originally to define the coded representation of high quality audio for storage media and a method for decoding high quality audio signals. Later the algorithms specified by MPEG were tested within the work of ITU-R for broadcasting applications and recommended for use in contribution, distribution, commentary, and emission channels [3]. Applications of MPEG-1 Audio included the digital compact cassette (DCC), digital audio and video broadcasting (DAB and DVB), transmission over ISDN, electronic music distribution (EMD), portable devices, etc. Arguably the commercially most successful embodiment of this standard is the widely adopted MP3 format.

M. Bosi (✉)
CCRMA, Stanford University, Stanford, CA USA
e-mail: MBOSI@STANFORD.EDU

L. Chiariglione (ed.), *The MPEG Representation of Digital Media*,
DOI 10.1007/978-1-4419-6184-6_6,

The initial goal for MPEG-2 Audio [4] was to define the multichannel extension to MPEG-1 (MPEG-2 BC, backwards compatible) and to define audio coding systems at lower sampling frequencies than MPEG-1, namely at 16, 22.5 and 24 kHz (MPEG-2 LSF). After a call for proposals in late 1993, work on a new approach to multichannel audio, the so-called MPEG-2 non-backwards compatible, NBC (later renamed MPEG Advanced Audio Coding, AAC), was started in 1994. The objective was to define a higher-quality multichannel standard than achievable with MPEG-1 extensions. MPEG-2 AAC is the core of the most advanced waveform-based audio coding technology (deployed in applications such as iTunes, etc.) and the basis for high-quality perceptual audio coding defined in MPEG-4 (see also next chapter).

From the technical point of view, perceptual audio coding stands at the cross road of a number of stimulating research and engineering application topics: from multi-rate signal processing to psychoacoustics modeling, from rate/distortion optimization problems to new types of audio quality assessments and implementation complexity issues. Just as basic digital signal processing horsepower became available to afford real-time implementations of more complex schemes, this novel field took off running and landed into everyone's hand less than two decades after the formulation of the first ideas.

In looking back at the history of perceptual audio coding development one can identify some remarkable milestones. Starting the eighties using simple adaptive differential pulse-code modulation (ADPCM) schemes, audio coding evolved rapidly with the adoption of pseudo-quadrature mirror filter (PQMF) [5, 6] based filterbanks and the exploitation of psychoacoustics models into more sophisticated systems by the end of the eighties. The introduction of the modified discrete cosine transform (MDCT) [7] allowed for the birth of high-quality transform-based perceptual audio coding at the beginning of the nineties and marked the consolidation of the basic building blocks for waveform audio coding as still currently in use in many coding schemes. Building upon these advances, new ways of looking at audio signals in terms of parametric representations of spectral content and spatial attributes were developed allowing for drastic reduction in date rates. Examples of these new approaches include parametric sinusoidal coding, spectral band replication, and parametric stereo/surround coding (see also the next two chapters of this book).

If one looks at the changes in data rates over time, we went from 256 kb/s for a stereo signal in 1992 using MPEG-1 Layer II (see the following sections in this chapter) down in recent years to 64 kb/s for 5-channel systems using MPEG Surround (see Chap. 7 of this book). Interestingly, while the aggregate bandwidth for media transmission and the storage capacity of personal devices has been increasing every year, this doesn't seem to have dampened the demand for audio compression. As a warning, especially for the younger generations, we should always be aware of the tradeoffs between quality and data rate so that our media choices allow for the optimal music experience with the available software/hardware. It is the hope that this chapter will provide enough information about the basic technology behind current perceptual audio coding technologies for the reader to exercise the his/her best choices at any time.

6.2 Design Criteria and Basic Tools for Audio Coding

In this section the basic tools employed in MPEG audio coding are described along with design criteria for these tools. These tools are still currently in use in waveform coding and represent the backbone of some of the most successful perceptual audio coders in the marketplace such as, for example, the MP3, MPEG AAC, and AC-3 coders [2, 4, 12, 16].

6.2.1 Design Criteria

In audio coding, as well as in speech and video coding, the input signal is a sampled and quantized version of the original signal. The decoded output signal is a close, albeit delayed, replica of the original signal where some distortion may be present depending on the data rate. The coding design goal is to obtain an output signal with the minimal perceived distortion with respect the original signal that can be achieved with a given data rate.

In classical source coding the signal is coded by exploiting its statistical properties and the emphasis is on the removal of redundancies in its representation. In some cases a model of the sound source – e.g. the vocal tract in speech coding – is developed as part of the coding process. A precise source model allows for very high compression ratios. Unfortunately, for general audio coding it is not feasible to successfully apply this approach since the coder must be able to handle a wide variety of signals with no pre-defined methods of generation. In addition to the removal of redundancies (like in classical source coding), perceptual audio coding emphasizes the removal of irrelevancies in the signal. In this context, irrelevancies are defined as parts of the signal representation whose removal is not perceived by the auditory system. In these coders, distortion in the output signal at the given data rate is allocated so that its audibility is minimized.

The foremost criterion in the design of an audio coder is to achieve the highest quality at a given data rate, where the audio quality is not defined in terms of traditional objective measurements such as signal to noise ratio (SNR), total harmonic distortion (THD), etc., but it is instead linked to perceived differences between the coder output and the original signal (see also Chap. 10 for details of quality assessment for audio coders). For example, one may argue based on the test results published in literature, see for example [8], that the most suitable coder for high quality audio applications is MPEG AAC at 64 kb/s per channel (while some of the coding schemes described later in this book perform better at much lower data rates, AAC seems to provide highest quality, see for example [9]). For many applications high fidelity audio is a strict requirement, therefore setting a strong lower limit on allowed data rates.

Another important criterion is to minimize both the computational complexity and memory storage requirements of the system at a given quality/data rate. While complexity issues may seem irrelevant due to the ever-increasing computing horsepower available, it is still an important issue since we tend use the horsepower to fit

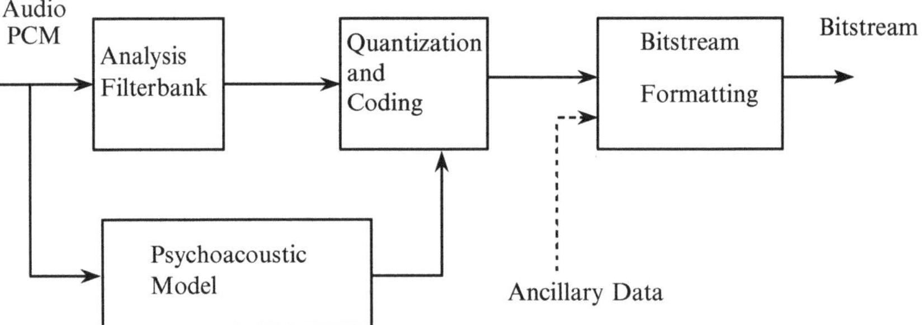

Fig. 6.1 Basic building blocks of the audio encoder

more capabilities in smaller and smaller devices. Finally, algorithmic delay, error resilience, editability, etc., are other requirements/criteria addressed in more recent audio coders (see next chapters for more details).

6.2.2 Basic Building Blocks

The typical basic building blocks of a perceptual audio encoder are shown in Fig. 6.1. The input signal channel is processed in parallel by an analysis filterbank and a psychoacoustic model. The analysis filterbank, typically a PQMF, MDCT, or hybrid filterbank (see next sections), decomposes the input signal into its spectral components. For each analysis block of time samples, the signal representation is converted from a time domain representation into the frequency domain. Given the statistical properties of audio signals, i.e., their quasi-periodicity, this conversion allows for a more compact representation of the signal and easier removal of redundancies. In addition, frequency representation of the signal provides a more natural framework for exploiting psychoacoustics models, so it also allows for easier removal of redundancies.

The goal of the psychoacoustic model is to estimate allowed levels of undetectable distortion by computing the masking curves (see also next sections) relative to the input signal. For each block of input samples, the psychoacoustic model may compute its own frequency representation via a fast Fourier transform (FFT). Based on the signal energy found in each frequency band and masking models derived from empirical data, masked curves and, specifically, time-dependent signal to mask ratios (SMRs) for each frequency band of the signal are derived. The SMR values represent the difference between the signal energy and its masking properties (where higher SMR values indicate lower masking).

The SMR values from the psychoacoustic model guide the quantization and coding stage where the spectral components of the signal are quantized. Quantization bits are allocated with the aim of keeping the quantization distortion introduced in this stage below the masked thresholds. This means that the SNRs resulting from the quantized signal have higher values than the SMRs for each spectral band. Typically, the core of the quantization stage is represented by a scalar midtread quantizer in conjunction with simple block or power law companding followed in some cases by additional noiseless coding such as Huffman coding (see also next sections).

In addition to the basic single-channel tools described above, techniques such as mid/sum (M/S) stereo coding (see, for example, [10]), intensity stereo coding [11], and coupling [12] are often used. These techniques aim to the further reduction of redundancies and irrelevancies between multiple audio channels. A better representation of a stereo signal in the M/S stereo coding scheme is afforded by adaptively coding either the left and right channel separately or the their sums/differences. This can be done on a frequency band-by-band basis or over the entire signal content. The added efficiency afforded by the M/S stereo technique is mainly due to spatial redundancy removal due to mono-like signal input and it is heavily dependent on the characteristics of the stereo input signals. M/S stereo coding is invertible and greatly relevant in high quality audio applications.

Intensity stereo coding is based on the observation that for high frequencies the energy envelope of the signal is the most important cue exploited for spatial perception. Based on this assumption, at high frequency, only the combination of the stereo channels is coded plus scale factors for the right and the left channels on a band-by-band basis. The main emphasis of intensity stereo coding is on the reduction of spatial irrelevancies. While the energy of the stereo signal is preserved, there is a potential loss of spatial information resulting in distortion. In general, however, the loss of spatial information is considered less annoying than other coding artefacts at low data rates. Coupling may be considered as an extension of intensity stereo coding to multichannel coding. A coupling channel is the equivalent of an n-channel combination of a stereo intensity channel. Instead of n separate channels, at high frequencies only one channel plus scale factors for each of the separate n channels are computed on a band-by-band basis.

The last stage in the encoder is the bitstream formatting stage. The bitstream syntax is defined in detail in the standards specifications and also test vectors are provided in the compliance part of the standard to ensure interoperability. Typically, the audio bitstream consists of the quantized and coded spectral coefficients of the audio signal plus control parameters, such as bit allocation parameters, and optional ancillary data.

A central component in the design of perceptual audio coders is the assessment of the performance not only for the final coder but for each of the coder stages. Many decisions and judgment calls in the practical design are determined, in addition to theoretical principles and implementation constraints, by quality assessments of the prototypes. While traditional quality measurements such as SNR and THD are meaningless for perceptual audio coders, objective perceptual measurements such as perceptual evaluation of audio quality (PEAQ) [13] are sometimes used. Ultimately, however, listening tests will provide the final saying as to whether the technology performance satisfies the requirements. Familiarity with coding artifacts and the ability to perform listening tests (see also Chap. 10) are important tools in developing audio coding systems.

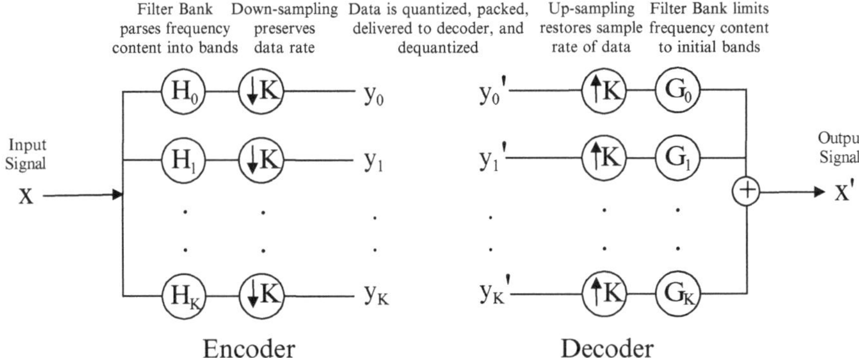

Fig. 6.2 Overview of the analysis/synthesis process (From Bosi and Goldberg [14])

6.2.3 *Filterbanks*

The choice of the analysis/synthesis filterbank in the encoding/decoding process is key in setting up the system coding efficiency. By determining the time/frequency resolution of the audio representation, the features of the filterbank control the level of effectiveness in the exploitation of psychoacoustics models and removal of redundancies in the input signal and, ultimately, in the quality/rate of the coder.

In Fig. 6.2 the basic block diagram of the analysis and synthesis filterbanks is shown. Specifically a static, critically-sampled, K-frequency channel analysis/synthesis filterbank is shown. The number of frequency channels (as well as the window/prototype filter, see below, and sampling rate) provides the basis for the frequency resolution of the coding system. The basic idea is to use the analysis filterbank to parse the signal into K "separate" bands of frequencies. Quantization with a limited number of bits is then applied to each band based on the signal strength and masking in that band. As a result, the quantization noise is confined in spectral regions where it is least audible. The quantized signal serves then as the input to the decoder where the coded signal is dequantized and the frequency bands are combined to restore the full frequency content of the signal. An additional synthesis filterbank is needed in the decoder to make sure that each band's signal remains limited to its appropriate frequency range before the bands are added together to create the decoded signal.

In MPEG audio coding schemes we find two main types of filterbanks: low frequency-resolution and high-frequency resolution filterbanks, where higher frequency resolution filterbanks in general afford larger coding gains for most signals. The first type, with 32 frequency channels, is typically implemented using a PQMF and the resulting coding system is traditionally described as a "subband coder". The PQMF is implemented by designing a narrow-band low-pass filter (prototype filter) that satisfies certain conditions that greatly reduce aliasing errors, and then cosine-modulating the prototype filter to desired locations in the frequency domain to isolate different components of the input signal.

High resolution filterbanks are implemented in MPEG audio coding using an MDCT. Before applying an MDCT, the input signal is first "windowed" or multiplied by a window function in order to minimize blocking artifacts. The high resolution MPEG filterbanks use an MDCT (1,024 frequency channels) as in the AAC scheme or a hybrid filterbank (576 frequency channels) as in the MP3 scheme where a PQMF is cascaded with an MDCT. The systems based on these filterbanks are traditionally described as "transform coders". It should be recognized, however, that there is no fundamental difference between the PQMF and the MDCT approach, see Ref. [15], making the "traditional" distinction between subband and transform coders somewhat superfluous. MDCT-based filterbanks are currently in use in many coding applications (see for example, AAC, AC-3, etc.).

The window function in an MDCT performs a role analogous to that of the prototype filter in the QMF approach in that it controls how the frequency spectrum in a single frequency channel is constrained to a range around the desired center frequency. In addition to the number of frequency channels, the prototype filter (PQMF) and the window (MDCT) characteristics are also key factors in the performance of the coding system.

While not a strict requirement, perfect reconstruction filterbanks (defined as filterbanks where the output of the synthesis filterbank is a delayed replica of the input to the analysis filterbank when no quantization noise is added) are desirable. The perfect reconstruction characteristic simplifies the design of the coding system but imposes further constraints on the window/prototype filter chosen. Filterbanks currently in use for perceptual audio coding are either perfect reconstruction (MDCT) or near perfect reconstruction (PQMF). In practice, sine windows or normalized windows such as the Kaiser-Bessel derived (KBD) [16] windows are used in MDCT-based filterbanks because they satisfy perfect reconstruction requirements for the MDCT.

In general, time-varying filterbanks are highly desirable in audio coding because they allow the tradeoff between time and frequency resolution to be dynamically adapted to the characteristics of the input signal. To achieve this goal a number of different approaches have been proposed in literature. The method currently used in the MDCT-based MPEG coders was first proposed by Edler [17]. Figure 6.3 shows the transition window and window sequences from long blocks for high frequency resolution to short blocks for high time resolution. The frequency/time resolution of the filterbank is dynamically changed while preserving the overall perfect reconstruction properties of the MDCT. This dynamic adjustment allows for better representation of both steady state and transient signals than a static filterbank which must find a middle ground to accommodate both types of signals.

6.2.3.1 Hybrid Filterbank

A hybrid filterbank consists of a cascade of different types of filterbanks. In MPEG-1 Layer III and in MPEG-2 Layer III, each of the 32 frequency channels of the PQMF is cascaded with and MDCT with 18 frequency channels for a total of 576 frequency

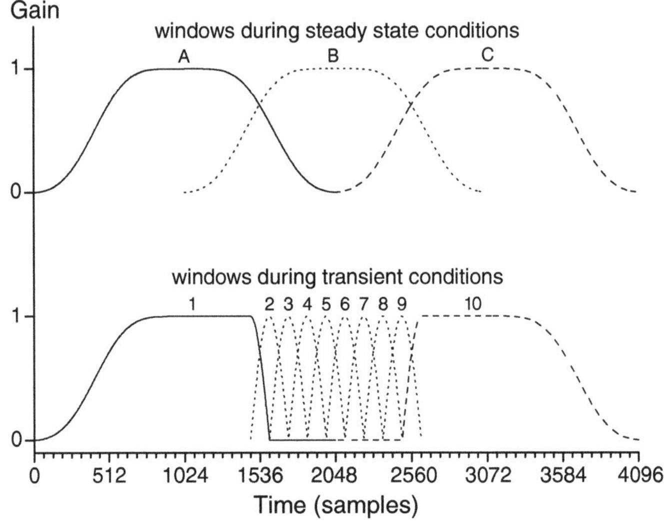

Fig. 6.3 Window sequence (lower part of the figure) from high frequency resolution to high time resolution and back to high frequency resolution in the AAC coding scheme; in the upper part of the figure a KDB window sequence used in steady state conditions is shown for comparison (From Bosi et al. [18])

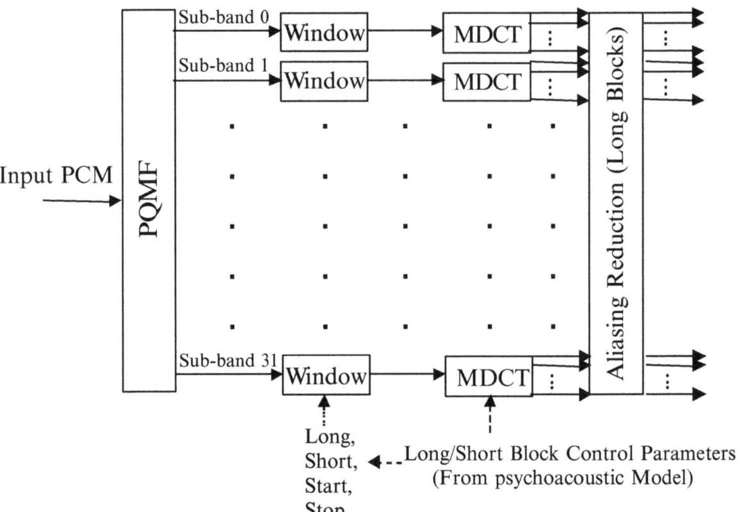

Fig. 6.4 MPEG Layer III analysis filterbank structure

channels (steady state) or 6 frequency channels for a total of 192 frequency channels (transients), see Fig. 6.4. Since the frequency response of the hybrid filterbank is the product of the two separate (PQMF and MDCT) responses, potential aliasing problems of the PQMF (crosstalk between subbands) are "emphasized" by the MDCT resulting in possible spurious peaks within stopbands. A solution to mitigate this problem was proposed by Edler [19] and integrated into the standard specifications [2].

6.2.4 Perceptual Models

A model of sound perception is at the core of perceptual audio coding. The main goal of the perceptual model in audio coding is to provide an accurate estimate, according to the time/frequency resolution of the coding system, of the quantization noise which is undetectable.

Work on the perception of sound can be traced back to the 1860s when Helmholtz suggested that the inner ear acts as a spectral analyzer. In 1940, Fletcher published a seminal paper [22] in which the auditory system was modeled as a "spectrum analyzer" with independent band-pass filters with frequency-dependent bandwidths or critical bands. A critical band was defined from experimental data as the bandwidth of a narrow-band noise which effectively masks a tone and apparently corresponds to a constant distance along the basilar membrane in the cochlea (rather than being a constant frequency interval). The Bark scale is a scale whose unit, the Bark, represents a critical bandwidth.

When a listener is presented with a loud sound the perception of other weaker sounds may change from the perception recorded in quiet. "Masking" occurs when a softer signal (maskee) becomes inaudible in the presence of a louder signal (masker) nearby in frequency (simultaneous masking) or time (temporal masking). In general masking depends both on the frequency/temporal characteristics of the masker and maskee and also shows non-linear behavior with respect to their intensities, see Ref. [20].

Experimental data [20] show that the masking effects of a certain masker spread through the spectrum beyond the critical band centered on a masker's frequency. The spreading shape, when represented on a Bark scale, appears to be independent of the center frequency of the masker. For this reason, masking models are typically characterized in the Bark scale rather than in frequency units. Experimental data from Ref. [20] for narrow-band noise masking tones show that the slope of these curves towards lower frequencies is constant at about 27 dB per Bark while the slope towards higher frequencies is dependent on the level of the masker for high levels (i.e., above 40 dB) and exhibits a non-linear, flattening effect for higher levels of the masker.

The noise or tone-like characteristics of the masker/maskee also play an important role. "Asymmetry" of masking was first recognized by Hellman [21] where a noise-like masker exhibits greater masking ability than a tone-like masker, i.e., noise components show higher masking thresholds than tone components at the same center frequency. Differential treatment of noise-like and tonal maskers is typically carried out in the psychoacoustic models used by perceptual coders.

The hearing threshold, or threshold in quiet, represents the lowest sound level of a pure tone that is just audible. This curve is extremely important in audio coding since frequency components that fall below this curve are deemed irrelevant to the representation of the audio signal. The threshold in quiet has been measured extensively, see, for example, [20]. It shows high levels at low (below 500 Hz) and high (above 10–15 kHz) frequencies, remaining constant at around 0 dB in the mid-range frequencies (between 500 Hz and 2 kHz) or dropping even below zero (between 2 and 5 kHz) for listeners with "good hearing". This implies that not all frequency levels are perceived in the same fashion. While the mid-range frequencies components are most relevant, very low and very high frequency components require less accuracy in their representation.

In order to dynamically create masked thresholds, i.e. thresholds below whose values signals are not audible, tonal and noise-like maskers are identified and their levels determined based on the computation of an FFT. Then, the spreading function and models of the asymmetry of masking are utilized to compute the masking threshold from each masker as a function of frequency. Finally, the effects of all masking thresholds and the threshold in quiet are combined to derive the overall masked threshold for each input block.

The masking thresholds and the threshold in quiet can be combined together in different fashions when computing the overall masked threshold. One way is to sum the relative intensities from each separate threshold; another is to use the highest threshold at a given frequency; and sometimes the summation of loudness of each threshold is carried out see, for example, PEAQ [13]. Decisions between these and other aspects of psychoacoustic models are typically "tuned" to a specific coder implementation.

It should be noted that, while simultaneous masking is at the heart of perceptual modeling for audio coding, temporal masking also plays a role [20], most importantly near transients in the signal. In temporal masking, loud signals may mask softer ones before or after them. However, temporal masking is asymmetric in that pre-masking (masking signals that precede) tends to be effective for shorter time scales than is post-masking (masking signals that follow). If coding artifacts are spread in time preceding an attack in the signal (which sometimes occurs due to suboptimal time resolution of the coding system- see for example low bitrate coded versions of the EBU SQAM castanet excerpt [23]) an audible "unmasked" distortion referred to as "pre-echo" may result.

6.2.4.1 MPEG Psychoacoustic Model 1

The role of MPEG is to define the digital representation of the audio signal that can be unambiguously decoded by a conforming decoder; therefore the psychoacoustic models used at the encoder stage are outside the scope of the standard and are described in an informative annex. This approach allows for changes and improvements in this part of the coding system as long as there is compliance with the bitstream syntax and the decoder structure defined in the normative part of the standard.

The input to the psychoacoustic model stage is the time representation of the audio signal over a certain time interval and the output is a set of signal to mask ratios (SMRs) for the main data path frequency partitions. The SMR values, based on the masked curves and the signal spectral energy computed for each input block, give an estimate of the audible portion of the signal for each block. This information is then used in the quantization and coding stage in order to make sure that the noise introduced by quantization falls below the masked thresholds.

Two main approaches are described in the MPEG informative part: Psychoacoustic Model 1, typically applied to Layers I and II (see also following sections), and Psychoacoustic Model 2, typically applied to Layer III and AAC.

In the first stage of Model 1 the input block is windowed with a Hanning window and transformed into the frequency domain using an FFT. For each frequency channel of the main data path PQMF the signal sound pressure level (SPL) is computed, where the signal level is normalized so that the level of a full scale input sine wave is equal to 96 dB when integrated in frequency. Next tonal and non-tonal maskers are identified in the signal spectrum and, in order to decrease computational complexity, irrelevant maskers are eliminated in a process called "masker decimation".

In order to create the masking curves, the tonal and non-tonal maskers' SPL is computed and "spread" in frequency by applying a spreading function. The spreading function used in Model 1 has a two-piece slope for both higher and lower frequencies and seems to be consistent with data for tones masking tones [20]. The peak masking level in Model 1 depends on the tonal versus noise-like characteristics and on the center frequency of the maskers.

The global masked threshold is then computed as the intensity sum of the relative individual thresholds and the threshold in quiet. Note that the threshold in quiet level is aligned to the signal representation so that the minimum of the threshold in quiet corresponds to a level of 0 dB.

The last step in the psychoacoustic model is to compute the SMR values for each frequency band in the main data path of the coder. For Model 1, since the PQMF spans more than one critical band, the minimum value of the global masked threshold for the group of critical bands covered by one PQMF frequency band is subtracted from the value of signal level for that band to compute the SMR.

6.2.4.2 MPEG Psychoacoustic Model 2

Model 2's computation approach is substantially different from that of Model 1. In Model 2, two Hanning-windowed FFTs are applied to the input blocks where the centers of these FFTs correspond to the center of the first and second half of the main data path input block. Model 2 then uses the lower of the two calculated masked threshold to compute the SMRs. To calculate the masked thresholds, the FFT frequency lines are grouped into partitions whose width is roughly 1/3 of a critical band. The energy densities are summed over each partition to create a masker level for each partition. The masking curve is then computed by convolving the Model 2 spreading function (see Ref. [14]) with the masker energy values and down-shifting the result based on an assessment of how tonal or noise-like the aggregate masking is at each frequency location. (This is easily implementable because the spreading function used in Model 2 is not level-dependent. Also, the convolution corresponds to combining the maskers by summation of the intensities corresponding to the "spread" energy of the maskers.) The value of the down-shift is linearly interpolated between 6 dB (noise-like maskers) and 29 dB (tonal maskers) for Layer III or 18 dB for AAC. The interpolation is carried out via a tonality index whose values range from zero for noise-like signals to one for tonal-like signals and which are computed separately based on the signal predictability [24]. For stereo signals the downshift is also compared with a frequency-dependent minimum value described in the standard that controls stereo unmasking affects (see also next sections) and the larger value is used for the down-shift. The masking levels for each partition are compared with the threshold in quiet and the larger is used in the computation of the SMR for that partition. For each partition the masking energy is evenly mapped to the frequency lines in that partition and subtracted from the energy level for each line in order to compute a set of SMR values for the main data path frequency bands.

In addition to the basic SMRs computation, Model 2 also controls the frequency/time resolution of the main data path filterbank. In order to achieve this, the model is calculated twice in parallel: once using a long block of 1,024 samples and once using a short block of 256 samples. Transients are detected based on a measure of "perceptual entropy" [25] and, in the presence of transients, the model can switch to short blocks. Perceptual entropy is an estimate of the minimum number of bits per block required to achieve transparency based on computed masked thresholds. Perceptual entropy values above a certain limit trigger the identification of transient like signals (as opposed to steady state signals) which in turn determines the selection of the appropriate block/window size for the main data path filterbank in order to adjust its time/frequency resolution to the input signal.

6.2.5 Quantization and Coding

The main data reduction tool in a perceptual coder is the quantization and coding stage. In this stage the frequency representation of the signal (output of the main data path filterbank) is quantized using different numbers of bits in different parts of the spectrum. The total bit pool is limited by the target data rate and the allocation is based on the SMR values computed in the psychoacoustic model. In order to more easily exploit psychoacoustic models, the quantizer is typically a scalar quantizer using "scale factor" bits to align the quantizer overload level with the signal and "mantissa" bits to carry out the scalar quantization. The input to the quantizer is the spectral representation of the input signal and the number of mantissa bits allocated.

Since quantization typically increases its signal-to-noise ratio (i.e., decreases its quantization noise level) by about 6 dB for each additional mantissa bit allocated, a simple bit allocation strategy for putting the quantization noise level below the masking level would be to assign nbits(b) mantissa bits to a frequency band b having an SMR value of SMR(b) as follows:

$$\text{nbits}(b) = \text{SMR}(b)/6.02\ \text{dB}$$

More generally, to respect the data rate constraint, bit allocation becomes a rate/distortion optimization problem which takes into account the total number of bits per block available as mantissa bits are allocated across the spectrum, see Ref. [14].

In MPEG Layers I and II a block companding method is utilized in which adjacent output samples from the PQMF sub-bands are normalized to a maximum absolute value. All values within this subblock of adjacent samples are scalar quantized with the same number of mantissa bits derived from the SMR values. In addition to the normalized values, the normalization values using 2 dB increments (factors of $2^{1/3}$) are coded for each subblock with a predetermined fixed number of scale factor bits. The bit allocation is implemented as an iterative process where, after initializing the process by setting all bit allocation codes to zero and assuming no bits are needed to transmit scale factors, in each iteration additional bits are allocated to the sub-band with the highest noise-to-mask ratio (NMR) until there are not enough additional bits left for the next iteration pass. The NMR for each sub-band is calculated as the difference between the SMR (calculated in the psychoacoustic model) and the signal-to-noise ratio (estimated from a table lookup based on the number of bits allocated to the sub-band). In Layer II bit allocations tables defined in the standard [2] are also utilized.

In MPEG Layer III and AAC a power law companding method is utilized where the output of the MDCT is raised to the ¾ power before quantizing with an SMR-determined number of bits. In addition, depending on the SMR values and the bit pool available, subblocks of adjacent MDCT output samples loosely following the critical bandwidth rate, may be amplified with a step size of 1.5 dB in order to contain the quantization noise below the masked curves. The "amplification" values, also referred to as scale factors, are coded with a predetermined fixed number of bits.

In addition to the quantization stage, a "coding" stage is present for Layers II and III, and AAC. For Layer II in the case of specific quantization step sizes (namely 3, 5, and 9) three consecutive samples are vector coded. This vector coding approach saves bits by allocating bits to enumerate the possible triplets rather than allocating bits individually to each quantized value. For Layer III and AAC, Huffman coding of the quantized frequency coefficients is utilized. For Layer III, three spectral regions are identified and each region is coded with a different set of Huffman code tables tuned for that region. At high frequencies an "all zeros" region is identified. A second region, containing values not exceeding magnitude one, is coded with 4-dimensional Huffman codes. Two Huffman tables are used for this region. The third region, or "big values" region, is further subdivided into three regions and is 2-dimensionally coded. Larger values in this region are accommodated by using escape mechanisms. 16 Huffman tables are defined in the standard for this region; furthermore, additional tables with different codeword length for the escape mechanism are also utilized. In AAC the Huffman encoding scheme is somewhat

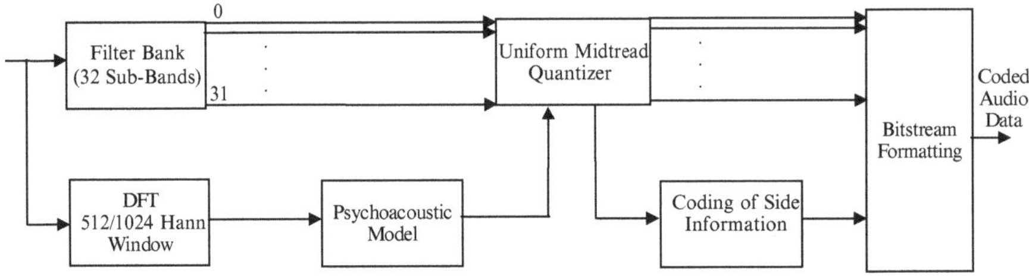

Fig. 6.5 Block diagram of MPEG Layers I and II (single channel mode)

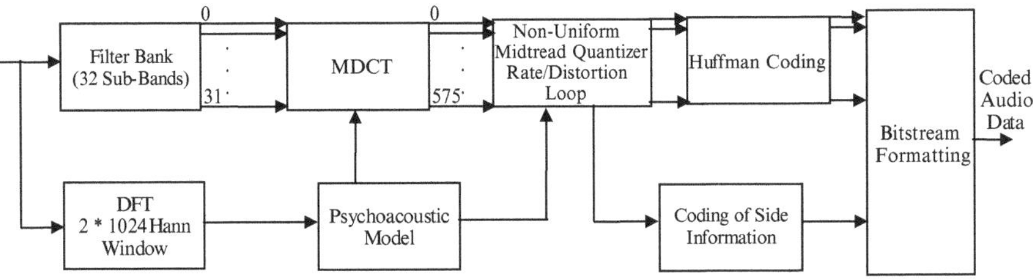

Fig. 6.6 Block diagram of MPEG Layer III (single channel mode)

similar to that of Layer III but more efficient. Firstly the spectral region subdivision is dynamic and typically varies from block to block, so that the number of bits needed to represent the full set of quantized spectral coefficients is minimized. The total number of tables for AAC is 12 (the zero table, four 4-dimensional tables, and seven 2-dimensional tables). Escape mechanisms as well as a "pulse method" for handling large magnitudes can be found in AAC. Finally, in addition to Huffman coding the spectral values, the scale factors are also Huffman encoded, significantly increasing the efficiency of the AAC coding stage with respect to Layer III.

Both in Layer III and AAC the spectral values are iteratively quantized based on the data rate available and the SMR values, the Huffman codes needed to represent the quantized spectral values are computed, and the resulting quantization noise is estimated. If the quantization noise in certain scale factor bands is above the masked threshold values calculated in the psychoacoustic model, the quantizer step size is modified for each of those scale factor bands by amplifying the relative scale factor. The quantization process is repeated until none of the scale factor bands have more than the allowed distortion, or the amplification for any band exceeds the maximum allowed value. In Figs. 6.5 and 6.6 the overall structure for MPEG Layers I, II, and III encoder is shown.

6.2.6 Stereo and Multichannel Coding

The goal of stereo and multichannel coding is to reduce the amount of spatial information in the signal representation without introducing audible artifacts or losing important spatial cues. Similarly to what is done for "intra-channel" coding, "inter-channel" coding is achieved by removing spatial redundancies and irrelevancies. Up to this point we have implicitly been considering the coding of a single monophonic signal; in this section we consider coding of two or more related audio channels.

Consider the case of a stereo signal where the left and right channel signals are almost equal. In such a case, a variable transformation into "mid" (M) and "side" (S) representation according to:

$$M = \frac{1}{2}(L+R) \qquad \qquad L = M + S$$
$$\text{or equivalently}$$
$$S = \frac{1}{2}(L-R) \qquad \qquad R = M - S$$

should lead to the S channel having much lower SPL than the M channel.[1] If a transformation from L/R to M/S leads to an S channel with much lower power than the M channel, then data rate savings can be gained by transmitting the full-range M signal along with the lower level S signal rather than transmitting the pair of full-range L and R channel signals. For this reason, it is common to compare the power of the S channel with the power of the M channel and, if the S channel has significantly less power in it, to convert from L/R to M/S representation before quantizing the spectral data. Again, the expectation is that a low-power S channel can be transmitted with far fewer bits than required for a full-bandwidth channel.

Consider also the case of a stereo signal where the perception of the left and right channels is dominated by a single source sound field X being transmitted in both channels:

$$L \approx aX \qquad \qquad L \approx aX \approx \left(\frac{a}{a+b}\right)(L+R) \approx \left(\frac{a}{a+b}\right)2M$$
$$\text{so that}$$
$$R \approx bX \qquad \qquad R \approx bX \approx \left(\frac{b}{a+b}\right)(L+R) \approx \left(\frac{b}{a+b}\right)2M$$

In this case, both the left and right channels can be approximated by a constant multiplier times the mid channel. This observation suggests an intermediate approach between passing the full L/R or M/S signals and neglecting the S channel entirely. Namely, it suggests that passing just the M channel and supplementing it with weights that parse out the total signal ($L+R=2M$) between the left and right channels may perceptually suffice to represent the signal. This is the idea behind the "intensity stereo coding" approach often applied at high frequencies at low bit rates – the high frequency content is converted from L/R representation into M/S representation but only the M signal is passed along with weighting factors for approximately

[1] Although this notation may suggest that we are blending the input signals in the time domain, this approach is usually carried out in the frequency domain over a frequency range in which the power of the spectral lines is similar in the two channels. This is because there is not much correlation between stereo channels in the time domain. In general, stereo redundancies can be more easily exploited for systems with high frequency resolution.

recovering the left and right channels from the *M* signal. Of course the true sound field is typically the sum of source signals at multiple locations so intensity stereo coding is throwing away much of the localization information present in the stereo signal. However, psychoacoustic studies have shown that our ability to localize sound accurately drops significantly at higher frequencies [26] so replacing the full localization at high frequencies with just the dominant component is viewed as an acceptable choice when the bit budget is very low.

Both the *M/S* transformation and the intensity stereo coding approaches have been extended to multichannel environments. For example, the *M/S* transformation can be applied to the signals from symmetrically placed speakers (e.g., to the *L/R* and L_s/R_s pairs in a surround environment with front *L/R* and rear surround L_s/R_s speaker pairs). Also, in such an environment pairs of channels that appear to have similar spectral shapes can be converted into a summed signal with weighting fractions analogous to the intensity stereo coding approach.

A first-cut approach to computing the masked threshold in a multichannel environment would be to identify maskers in each channel and use the full set of maskers to compute an overall masked threshold. This would be equivalent to allowing maskers in any channel to mask noise or soft signals in any other channel. If we assume that quantization noise is uncorrelated across channels then this masked threshold would be 3 dB above the noise that would be allowed in each channel and so the masked threshold used for each channel should be dropped by this amount. However, psychoacoustic experiments have also shown that masking can be even lower than this level due to our increased ability to detect signals binaurally, i.e., noise that would be inaudible in a mono signal can become unmasked in the corresponding stereo signal [26]. Depending on the relative phase of the maskers/maskees, the masked threshold can be depressed by a frequency-dependent amount often referred to the "binaural masking level difference" or BMLD. This implies that binaural phase differences can reduce masking over the level estimated based on monophonic signal modeling. Hence, the masked threshold computed using an approach allowing for masking from any channel should be further reduced based on BMLD data. In general, BMLD data show a large reduction in masking level (~15–20 dB) at low frequencies with very little reduction in masking at higher frequencies.

6.3 The MPEG-1 Standard

The input of the MPEG-1 audio encoder and the output of the decoder are compatible with existing PCM standards such as the CD and the digital audio tape, DAT, formats. MPEG-1 audio aimed to support 16-bit precision, one or two main channels, and sampling frequencies of 32, 44.1, and 48 kHz. The MPEG-1 standard describes a perceptual audio coding algorithm that is designed for general audio signals. There is no specific source model applied as, for example, in speech codecs. The audio signal is represented by its spectral components on a block-by-block basis and encoded exploiting perceptual models. The aim of the algorithm is to provide a perceptually lossless coding scheme. In addition to a monophonic mode for a single audio channel configuration, a dual monophonic mode for two independent channels is included. A stereo mode for stereophonic channels, which shares the available bit pool amongst the two channels but does not exploit any other spatial perceptual model, is also covered. Moreover, joint stereo modes that take advantage of correlation and irrelevancies between the stereo channels are described in the standard. Nominal data rates vary between 32 and 224 kb/s per channel allowing for compression ratios ranging from 2.7 to 24:1 depending on the sampling rate, although MPEG-1 is mostly used between 64 and 192 kb/s per channel. In addition to the pre-defined data rates, a free format mode can support supplementary, fixed data rates.

MPEG-1 and 2 Audio specify three layers. The different layers offer increasing audio quality at slightly increased complexity. While Layers I and II share the basic structure of the encoding process having their roots in an earlier algorithm also known as MUSICAM [27], Layer III is substantially different. The Layer III algorithm was derived from ASPEC [28], where instead of the MDCT based filterbank a hybrid MDCT-PQMF filterbank is utilized. MPEG-1 Layer I is the simplest layer and it typically operates at data rates of 192 kb/s per channel. Layer II is of medium complexity and it typically employs data rates of 128 kb/s per channel. Layer III exhibits increased complexity but lower data rates than Layers I and II (below 128 kb/s per channel). A modification of the MPEG Layer III format at lower sampling frequencies gave origin to the ubiquitous MP3 file format.

The quality of the MPEG-1 audio standard was tested by extensive subjective listening tests during its development. The resulting data, see for example [29], showed that, under strictly controlled listening conditions, experts listeners were not able to distinguish between coded and original sequences with statistical significance at typical codec data rates.

6.3.1 Layers I and II Implementation

For Layers I and II, the time to frequency mapping is performed by applying a 512-tap, 32-band PQMF (see also the Sect. 2.3) to the main audio path data. The frequency resolution of Layers I and II PQMF at 48 kHz sampling rate is 750 Hz and time resolution is 0.66 ms. Notice that his suboptimal frequency resolution limits the coding efficiency of Layers I and II, since we know from psychoacoustics data that the critical bandwidths at low frequencies are <= 100 Hz. In this region a single main data path sub-band covers several critical bands, therefore the critical band with the highest SMR (lowest masking ability) dictates the number of quantization bits needed for the entire sub-band.

The frequency representation of the signal is scaled and then quantized with a uniform midtread quantizer (see also Sect. 6.4) whose precision is determined by the output of the psychoacoustic model. The scale factors, mapped to a look up table defined in the standard, cover a dynamic range of 120 dB and are coded with 6 bits. Typically, one scale factor for every 12 consecutive samples or "granule" per PQMF band is derived. The bitstream is organized in data "frames".

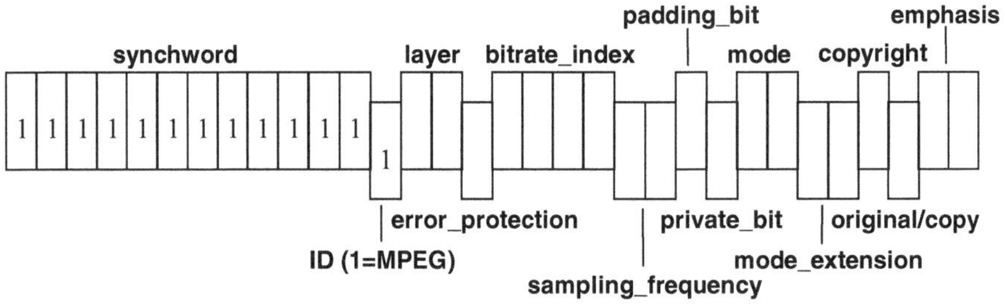

Fig. 6.7 MPEG-1 audio header

LAYER I	Header (32)	CRC (0,16)	Bit Allocation (128-256)	Scale Factors (0-384)	Samples	Ancillary Data

LAYER II	Header (32)	CRC (0,16)	Bit Allocation (26-188)	SCFSI (0-120)	Scale Factors (0-1080)	Samples	Ancillary Data

LAYER III	Header (32)	CRC (0,16)	Side Information (130-246)	Main Data (May start at a previous frame)

Fig. 6.8 MPEG-1 audio frame format

The data frame basic structure consists of one (Layer I) or three (Layer II) granules (i.e., 12 or 36 consecutive quantized samples for each PQMF band). For Layer II a scale factor select information (SCFSI), coded with two bits per subband, determines whether one, two or three scale factors are transmitted for the three consecutive granule in the frame. Each data frame is preceded by a 32-bit header, CRC, bit allocation, scale factors, and other control parameters (see also Figs. 6.7 and 6.8).

6.3.2 *Layer III Implementation*

For Layer III the output of the PQMF is fed to an MDCT stage. It should be noted that the Layer III filter bank is not static as in Layers I and II, but it is signal adaptive. The frequency resolution for Layer III at 48 kHz sampling rate is 41.66 Hz and time resolution is 4 ms. Notice that Layer III hybrid filter bank provides much higher frequency resolution than the Layers I and II PQMF. The time resolution, however, is decreased. The decreased time resolution renders Layer III more prone to pre-echo.

A number of measures to reduce pre-echo are incorporated in Layer III including a detection mechanism in the psychoacoustic model and the ability to "borrow" bits from the bit reservoir in addition to block switching. The inherent filter bank structure with long impulse response of $384+512=896$ samples even in the short

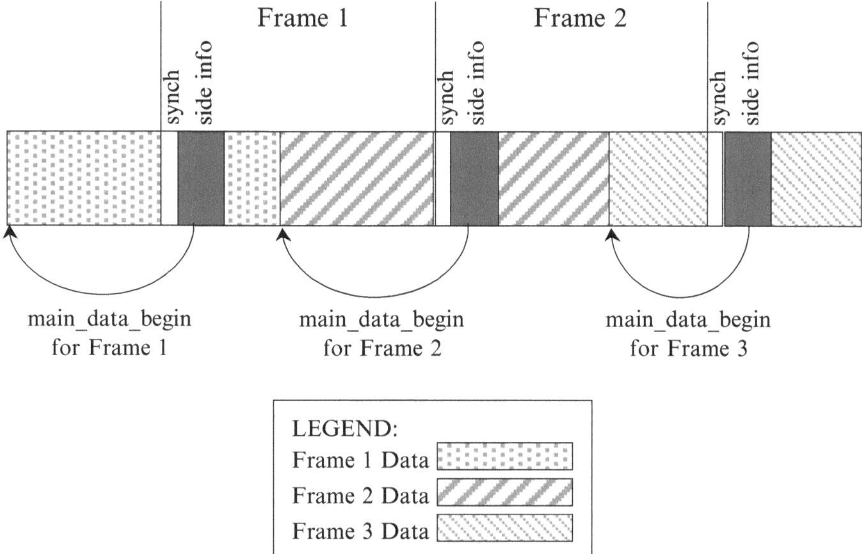

Fig. 6.9 Example of Layer III frame structure (From [ISO/IEC 11172–3])

block mode, however, makes the encoding of transients a challenge for Layer III. The output of this hybrid filter bank is scaled and then non-uniformly quantized with a midtread quantizer.

Noiseless coding is also applied in Layer III. In an iterative loop that performs the synthesis of the Huffman-encoded, quantized signal and compares its relative error levels with the masked thresholds levels, the quantizer step size is calculated for each spectral region.

The data frame basic structure consists of two granules of 576 samples (i.e., 18 quantized samples for each PQMF band). In addition, a bit reservoir method is implemented in Layer III. In this fashion, the bitstream formatting routine is designed to support a locally-variable data rate. This mechanism not only allows to respond to local variations in the bit demand, but also is utilized to mitigate possible pre-echo effects. Although the number of bits employed to code a frame of audio data is no longer constant, its long term average is. The deviation from the target data rate is always negative, i.e., the channel capacity is never exceeded. The maximum deviation from the target bit rate is fixed by the size of the maximum allowed delay in the decoder. This value is fixed in the standard by defining a code buffer limited to 7,680 bits. As shown in Fig. 6.9, the first element in the Layer III audio data bitstream is a 9-bit pointer (main_data_begin) which indicates the location of the starting byte of the audio data for that frame.

In MPEG-1 Audio a combination of M/S (broadband) and intensity stereo coding is supported. In all layers the audio data together with the side information such as bit allocation and control parameters are multiplexed with the optional ancillary data and then stored or transmitted.

6.4 The MPEG-2 Standard

Motivated by the increasing number of low data rate applications over the Internet, the goal of MPEG-2 LSF was to achieve MPEG-1 or better audio quality at lower data rates. One way to achieve this goal without requiring major modifications in the MPEG-1 system was to decrease the sampling rate of the audio signals which is the approach taken by MPEG-2 LSF. Instead of the 48, 44.1, and 32 kHz sampling rates seen in MPEG-1, the sampling rates for MPEG-2 LSF are 24, 22.05, and 16 kHz. Of course, reducing the sampling rate by a factor of 2 also reduces the audio bandwidth by a factor of 2. This loss in high frequency content was deemed an acceptable compromise for some target applications in order to reduce the data rate by a factor of 2. The MPEG-2 LSF coder and its bitstream have a very similar structure to that of MPEG-1 Audio. The three audio "layers" are defined almost identically to the layers of MPEG-1 Audio. The bitstream header (see Fig. 6.7) for MPEG-2 LSF differs from that of MPEG-1 solely in the settings of the 13th bit (called the "ID bit") which is set equal to zero for MPEG-2 LSF while it was equal to one for MPEG-1. The bit setting for MPEG-2 LSF signals the use of different tables for sampling rates and target data rates. The main difference in the MPEG-2 LSF implementation is the set of allowed sampling rate and data rate pairings in its operation. An additional difference is the adaptation of the psychoacoustic parameters for the lower sampling frequencies. The frame size is reduced from 1,152 samples to 576 samples in Layer III to make the audio frame more manageable for packetizing in internet applications. The resulting data rates in MPEG-2 LSF are lower than the data rates for MPEG-1. The nominal operating rates for MPEG-1 are from 32 kb/s up to 224 kb/s but good quality is usually found around 128 kb/s per channel. In contrast, MPEG-2 LSF typically produces adequate quality for various applications throughout its operating range from 32 to 128 kb/s per channel for Layer I, and from 8 to 80 kb/s per channel for Layers II and III.

6.4.1 The MP3 Format

The decrease in data rates, especially for Layer III, made MPEG-2 LSF useful for Internet applications. This led to the creation of even lower sampling rate modification for Layer III, namely 12, 11.025, and 8 kHz and the creation of the MP3 format. To allow decoders to work with the same bitstream format as used in MPEG-1 Audio and MPEG-2 LSF, the final bit from the header synchword is removed and merged with the ID bit to become a 2-bit MP3 ID code that identifies the bitstream format. The ID codes are [00] for MP3, [11] for MPEG-1, [10] for MPEG-2 LSF, and [01] reserved for future extensions. Using a [1] as the first bit of the two-bit ID code bit for the prior formats leads to compatibility with the MPEG-1 and MPEG-2 formats because that bit would have been [1] as part of the synchword. This so-called "MP3 file format" is typically implemented using Layer III. The use of MP3 files for sharing audio over the Internet has spread so widely that it has become the de facto standard for Internet and portable devices audio.

6.4.2 MPEG-2 BC Multichannel Audio Coding

MPEG-2 multichannel audio coding provides a multichannel extension to MPEG-1 Audio which is backwards compatible (BC) with the MPEG-1 bitstream. (Backwards compatibility implies that an MPEG-1 decoder is able to decode MPEG-2 encoded bitstreams.) The audio sampling rates in MPEG-2 BC for the main channels are the same as in MPEG-1 but sample precision can be increased to 24 bits and the channel configuration supports up to five full-bandwidth main audio channels plus a low frequency enhancement (LFE) channel whose bandwidth is less than 200 Hz (the "0.1" channel). The sampling frequency of the LFE channel corresponds to the sampling frequency of the main channels divided by 96. The data rates supported run from 32 kb/s up to 1.13 Mb/s.

The 5.1-channel audio configuration is often referred to as the 3/2/.1 configuration, since three loudspeakers are typically placed in front of the listener, two are in the side/rear, and an LFE speaker is included. This arrangement is described in detail in the ITU-R recommendation BS.775-1 [30]. According to the ITU-R specifications, the five full-bandwidth loudspeakers are placed on the circumference of a circle centered on the reference listening position. Three front loudspeakers are placed at angles from the listener axis of −30° (left channel, L), +30° (right channel, R), and 0° (center channel, C); the two surround loudspeakers are placed at angles between −100° and −120° (left surround channel, L_s) and +100° and +120° (right surround channel, R_s).

The LFE speaker is typically placed in the front, although its exact location is not specified in the ITU-R layout. The purpose of the LFE channel is to enable high-level sounds at frequencies below 200 Hz without overloading the main channels. In the cinema practice, this channel has 10 dB more headroom than the main channels.

In MPEG-2 BC matrixing techniques are applied to the multichannel data. In this approach, the left and right channels contain down-mixed multichannel information which approximates the full multichannel signal. In general, the advantage of this approach is that little additional data capacity is required for multichannel service. The disadvantage is that this approach greatly constrains the design of the multichannel coder and so it limits the quality of the multichannel reproduction for certain classes of signals at a given data rate. Since the MPEG-1 equivalent left and right channels need to carry a complete description of the MPEG-2 main multichannel signal, the five-channel information is down-mixed into two channels as follows, see also Ref. [31]:

$$L_o = c\left(L + aC + bL_s\right)$$
$$R_o = c\left(R + aC + bR_s\right)$$

where L represents the left channel, R the right channel, C the center channel, and L_s and R_s the left and right surround channels respectively. L_o and R_o represent the MPEG-1 compatible left and right channels. The down-mix coefficients are specified in the standard. For example, commonly used values for the down-mix coefficients are given by

$$a = b = \frac{1}{\sqrt{2}} \quad \text{and} \quad c = \frac{1}{1+\sqrt{2}}$$

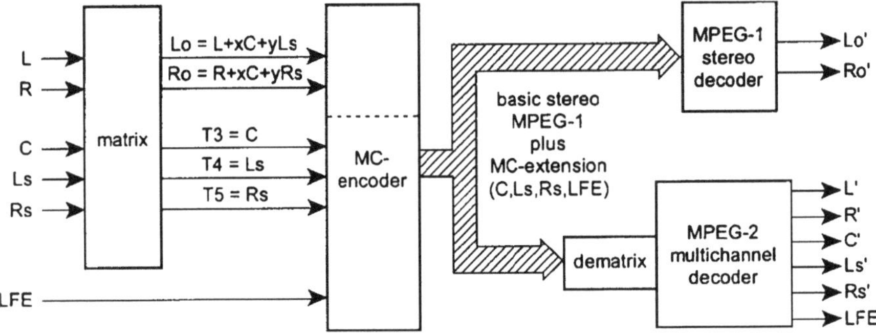

Fig. 6.10 MPEG-2 multichannel BC configuration (From Stoll [32])

While the basic MPEG-2 BC bitstream structure is essentially the same as in MPEG-1, additional channels, T_3, T_4, T_5 (corresponding to C, L_s, and R_s respectively see also Fig. 6.10) and LFE channels are stored along with the multichannel header as an optional extension bitstream in the ancillary data field.

In addition to the man audio channels, MPEG-2 BC supports up to seven multilingual channels.

6.4.3 MPEG-2 Advanced Audio Coding (AAC)

In 1994 the MPEG-2 Audio committee decided to define a higher quality multichannel standard than was achievable while requiring MPEG-1 backwards compatibility. The so called MPEG-2 non-backwards compatible audio (MPEG-2 NBC) standard, later renamed MPEG-2 Advanced Audio Coding (MPEG-2 AAC) [33] was finalized in 1997. AAC made use of all of the advanced audio coding techniques available at the time of its development to provide very high quality multichannel audio. Tests carried out in the fall of 1996 at BBC, UK, and NHK, Japan, showed that MPEG-2 AAC satisfies the ITU-R "transparency" quality requirements at 320 kb/s per five full-bandwidth channels (or lower according to the NHK data) [34]. The MPEG-2 AAC tools also constitute the kernel of the MPEG-4 general audio and scalable coder structure [35]. AAC sampling rates range from 8 kHz up to 96 kHz and AAC supports up to 48 channels (default configurations include 3/2/0.1, 3/1, …, etc. plus 16 program configurations), and up 24-bit precision. Data rates vary up to 576 kb/s per channel depending on the sampling rates, while there is no nominal limitation for the lowest data rates (AAC can be used at data rates as low as 8 kb/s per channel).

Three profiles are defined for AAC: Main, Low Complexity (LC), and Scalable Sampling Rate (SSR) Profiles. In the Main Profile all tools (with the exception of the preprocessing tool, see Fig. 6.11) may be used. In LC Profile, the most successful AAC profile in the marketplace, prediction and preprocessing tools are not used

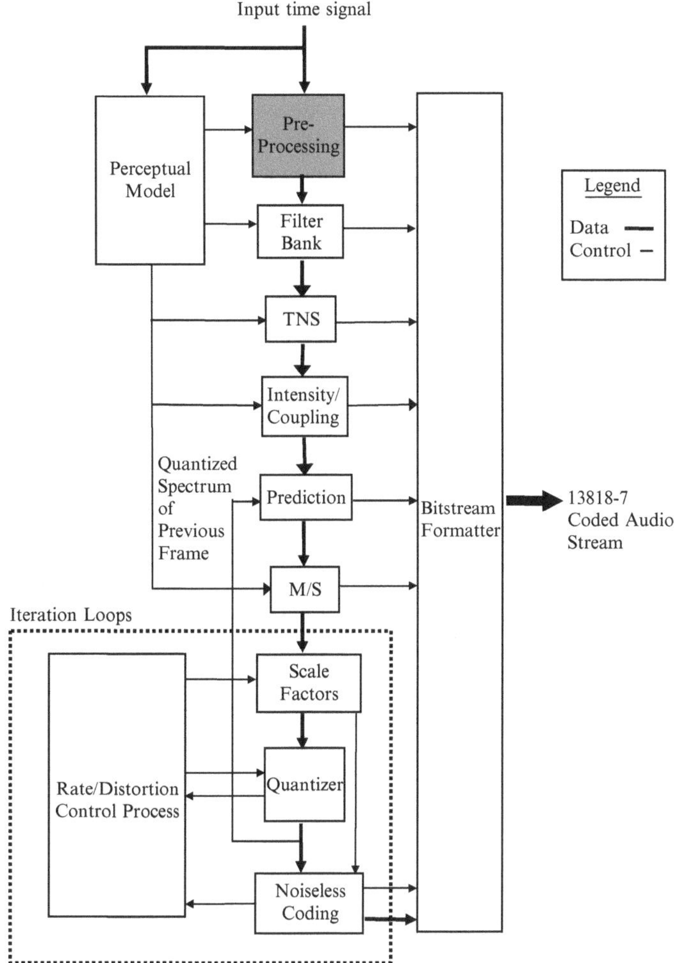

Fig. 6.11 MPEG-2 AAC encoder block diagram (From Bosi et al. [18])

and the temporal noise shaping, TNS, tool is of reduced complexity with respect to the Main Profile. The SSR Profile, which utilizes the preprocessing tool and reduced complexity TNS while not incorporating prediction, has lower complexity than Main and LC Profiles.

In Fig. 6.11 the AAC tool configuration is depicted. The preprocessing tool includes a time-domain gain control stage in order to mitigate pre-echo distortion and a 4-band conjugate quadrature filter (CQF) that parses the signal into four equally spaced frequency bands thus allowing for content and complexity scalability. The filterbank is a 1,024 or 256 frequency channel MDCT with a KBD or sine window. AAC filterbank exhibits higher frequency resolution than MPEG-1/2 Layers I, II and III which use 32/576 frequency lines vs. AAC's 1,024, improved window selectivity with the KBD window (versus the sine window for Layer III) and improved block switching (see Fig. 6.3) with window length of 2,048/256 (for comparison, the Layer III window length is 1,152/384) and impulse response of 5.3 ms at 48 kHz (versus Layer III 18.6 ms). The TNS tool was introduced in AAC in order to better shape quantization noise in the time-domain therefore mitigating the pre-echo distortion and improving the coder performance for transient signals especially for speech samples (in preliminary subjective tests conducted by the MPEG group TNS showed an improvement of over one point in the ITU-R impairment scale for the German Male Speaker excerpt from the SQAM disc). TNS, introduced by Herre and Johnston [36] is implemented by applying a filtering process to part of the spectral data.

AAC also introduces an improved joint stereo coding approach with respect to Layers I, II, and III. A generalized intensity coding technique or coupling is introduced as well as band-wise M/S coding. The prediction tool is included in the Main Profile only and is implemented as a second order backward adaptive predictor. For steady state signal, this tool helps to further reduce redundancies in the signal. The scale factor band width is roughly half a critical bandwidth providing enhanced band resolution with respect to Layer III. A more flexible Huffman coding scheme is applied as well Huffman coding of the scale factors thus increasing the overall efficiency of the coder. The AAC system has a very flexible bitstream syntax allowing for conditional components in order to minimize overhead. The data frame encompasses 1,024 quantized frequency samples but AAC allows for a variable number of bits per block adopting a bit reservoir mechanism similar to that of Layer III (AAC utilizes however variable-rate headers). Two bitstream layers are defined: the lower layer specifies the "raw" audio information while the higher specifies a specific audio transport mechanism. Since any one transport cannot be appropriate for all applications, the raw data layer is designed to be parsable on its own, and in fact is entirely sufficient for applications such as compression for computer storage devices.

A number of tests have been carried out to measure the performance of AAC. One of the most interesting tests because of its assessment in conjunction with the subjective evaluation of other state-of-the-art two-channel audio coders took place at the Communication Research Center, CRC, Ottawa, Canada [8] (see Fig. 6.12). MPEG-2 AAC was tested at 128 and 96 kb/s per stereo signal. Other codecs included in the tests were AC-3 [16] at 192, 160, 128, and 96 kb/s; MPEG Layer II at 192, 160, 128 kb/s; MPEG Layer III at 128 kb/s; and Lucent PAC [37] at 160, 128, 96, and 64 kb/s. At 128 kb/s AAC ranked the best among the codecs tested.

The MPEG-2 AAC coder is also at the core of the MPEG-4 General Audio scheme (see also Chap. 7) and is used as part the MPEG-4 AAC coder. Some of the most commercially successful applications of MPEG4 AAC are the MPEG-4 Audio Low Complexity Profile which is the MPEG-2 LC Profile with the addition of the perceptual noise substitution (PNS) tool [38] and the MPEG-4 High Efficiency Profile is the MPEG-2 LC Profile with the addition of the spectral band replication (SBR) tool [39].

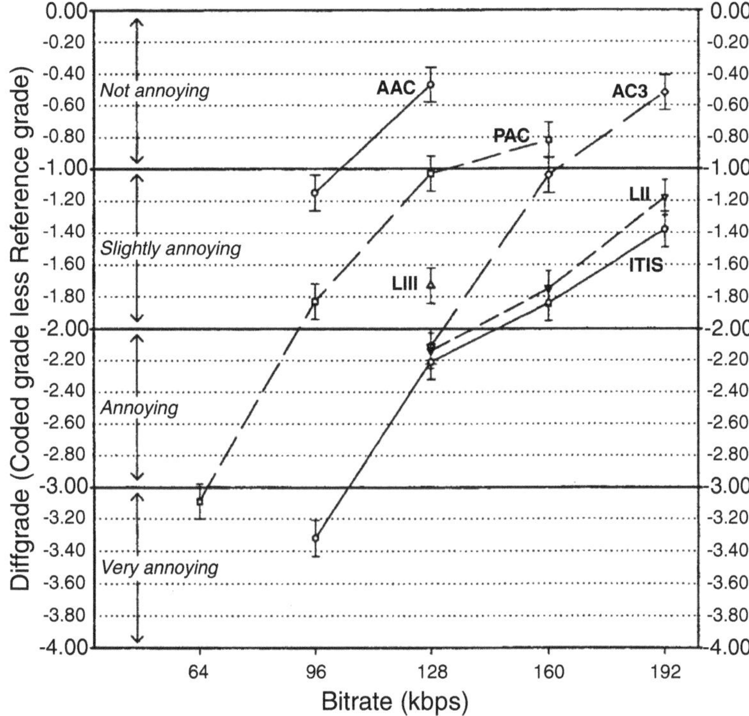

Fig. 6.12 Comparison of MPEG AAC overall quality with AC-3, MPEG layer II and layer III (From Soulodre et al. [8])

6.5 Summary

In this chapter we reviewed the basic features of the general purpose, perceptual audio coders described in the MPEG-1 and MPEG-2 audio standards. MPEG-1 Audio was the first international standard for high quality audio coding and opened the doors to a variety of applications. The widely successful MP3 and AAC coders are one of the most notable products of MPEG and highly influenced not only the technology but also largely enabled new ways of music consumption. One can look at the ubiquity of the portable devices and internet music distribution enabled by the MPEG Audio technology in the past few years to see the success of these technologies. More recent audio coding technologies that build upon and extend these important coding tools to continue and augment the success of MPEG Audio will be described in the next few chapters.

References

1. Schroeder M R, Atal B S, Hall JL (1979), Optimizing Digital Speech Coders by Exploiting Masking Properties of the Human Ear. J Acoust Soc Am, 66:1647–1652
2. ISO/IEC 11172–3 (1993), Information Technology, Coding of moving pictures and associated audio for digital storage media at up to about 1.5 Mbit/s, Part 3: Audio
3. ITU-R BS.1115 (1994), Low Bitrate Audio Coding. ITU, Geneva
4. ISO/IEC 13818–3 (1994–1997), Information Technology - Generic Coding of Moving Pictures and Associated Audio, Part 3: Audio
5. Nussbaumer H J (1981), Pseudo-QMF Filter Bank". IBM Tech Disclosure Bull, 24: 3081–3087
6. Rothweiler J H (1983), Polyphase Quadrature Filters - A new Subband Coding Technique. International Conference IEEE ASSP, Boston, 1280–1283
7. Princen J P, Johnson A, Bradley A B (1987), Subband/Transform Coding Using Filter Bank Designs Based on Time Domain Aliasing Cancellation. Proc of the ICASSP, 2161–2164
8. Soulodre G A, Grusec T, Lavoie M, Thibault L (1998), Subjective Evaluation of State-of-the Art Two-Channel Audio Codecs. J Audio Eng Soc, 46:164–177
9. ISO/IEC JTC 1/SC 29/WG 11 N7950 (2006), Performance of MPEG Surround Technology
10. J. D. Johnston and A. J. Ferreira (1992), Sum-Difference Stereo Transform Coding, Proc. ICASSP, pp. 569–571
11. vd Waal R G and Veldhuis R N J (1991), Subband Coding of Stereophonic Digital Audio Signals, Proc. ICASSP, pp. 3601 – 3604
12. Davies M (1993), The AC-3 Multichannel Coder, presented at the 95th AES Convention, New York, pre-print 3774
13. Thiede T, Treurniet W, Bitto R, Schmidmer C, Sporer T, Beerends J, Colomes C, Keyhl M, Stoll G, Brandenburg K, Feiten B (2000), PEAQ-The ITU Standard for Objective Measurement of Perceived Audio Quality. J Audio Eng Soc, 48:3–29
14. Bosi M, Goldberg R E (2003), Introduction to Digital Audio Coding and Standards. Springer, New York
15. Malvar H S (1990), Lapped transforms for efficient transform/sub-band coding. IEEE Transactions on Acoustics Speech and Signal Processing, 38:969 – 978
16. Fielder L, Bosi M, Davidson G, Davis M, Todd C, Vernon S (1996), AC-2 and AC-3: Low-Complexity Transform-Based Audio Coding. Collected Papers on Digital Audio Bit-Rate Reduction, Neil Gilchrist and Christer Grewin (ed), AES 54–72
17. Edler B (1989), Coding of Audio Signals with Overlapping Transform and Adaptive Window Shape (in German), Frequenz, 43:252–256
18. Bosi M, Brandenburg K, Quackenbush S, Fielder L, Akagiri K, Fuchs H, Dietz M, Herre J, Davidson G, Oikawa Y (1997), ISO/IEC MPEG-2 Advanced Audio Coding. J Audio Eng Soc, 45:789 – 812
19. Edler B (1992), Aliasing reduction in sub-bands of cascaded filter banks with decimation. Electronics Letters, 28:1104–1105
20. E. Zwicker E, Fastl H (1990), Psychoacoustics: Facts and Models. Springer-Verlag, Berlin
21. Hellman R (1972), Asymmetry of Masking Between Noise and Tone. Percep Psychphys, 11:241–246
22. Fletcher H (1940), Auditory Patterns. Rev Mod Phys, 12:47–55
23. EBU (1988), Tech 3253 - Sound Quality Assessment Material (SQAM). Tech Rep, European Broadcasting Union
24. K. Brandenburg K, Johnston JD (1990), Second Generation Perceptual Audio Coding: The Hybrid Coder. 88th AES Convention, Montreux
25. Johnston J D (1988), Estimation of Perceptual Entropy Using Noise Masking Criteria. Proc ICASSP, 2524–2527
26. Blauert J (1983), Spatial Hearing. MIT Press, Cambridge

27. Dehéry Y F, Stoll G, Kerkhof L vd (1991), MUSICAM Source Coding for Digital Sound. Symp Rec Broadcast Sessions, 612–617
28. Brandenburg K, Herre J, Johnston J D, Mahieux Y, Schroeder E F (1991), ASPEC-Adaptive Spectral Perceptual Entropy Coding of High Quality Music Signals. 90th AES Convention, 3011
29. Ryden T, Grewin C and Bergman S (1991), The SR report on the MPEG audio subjective listening tests in Stockholm April/May 1991, ISO/IEC JTC1/SC29/WG 11 MPEG 91/010
30. ITU-R BS.775-1 (1992–1994), Multichannel Stereophonic Sound System with and without Accompanying Picture
31. ten Kate W, Boers P, Maekivirta A, Kuusama J, Christensen K E, Soerensen E (1992), Matrixing of Bit-Rate Reduced Signals. Proc ICASSP, 2:205–208
32. Stoll G (1996), ISO-MPEG-2 Audio: A Generic Standard for the Coding of Two-Channel and Multichannel Sound. Gielchrist , Grewin (ed), Collected Papers on Digital Audio Bit-Rate Reduction, 43–53, AES 1996
33. ISO/IEC 13818–7 (1997), Information Technology - Generic Coding of Moving Pictures and Associated Audio, Part 7: Advanced Audio Coding
34. ISO/IEC JTC 1/SC 29/WG 11 N1420 (1996), Overview of the Report on the Formal Subjective Listening Tests of MPEG-2 AAC Multichannel Audio Coding
35. ISO/IEC 14496–3 (1999–2001), Information Technology – Coding of Audio Visual Objects, Part 3: Audio
36. Herre J, Johnston J D (1996), Enhancing the Performance of Perceptual Audio Coders by Using Temporal Noise Shaping (TNS). 101st AES Convention, 4384
37. Sinha D, Johnston J D, S. Dorward S, Quackenbush S R (1998), The Perceptual Audio Coder (PAC). The Digital Signal Processing Handbook, Madisetti, Williams (ed), CRC Press, 42.1-42.18
38. Herre J, Schulz D (1998), Extending the MPEG-4 AAC Codec by Perceptual Noise Substitution. 112th AES Convention, 4720
39. Dietz M, Liljeryd L, Kjoerling K, Kunz O (2002), Spectral Band Replication, a novel approach in audio coding. 112th AES Convention, 5553

Chapter 7
MPEG Audio Compression Advances

Schuyler Quackenbush

7.1 Introduction

Recent MPEG audio coding work continues the quest for ever greater compression efficiency. It uses the tools of the MPEG-1 and MPEG-2 perceptual coders and adds a number of new mechanisms to achieve greater signal compression, giving it the following "compression toolbox":

- Perceptually shaped quantization noise
- Parametric coding of high-band spectral components
- Parametric coding of the multi-channel of multi-object sound stage

MPEG-4 AAC is a minor extension of MPEG-2 AAC, and as such continues to exploit the mechanism of perceptually shaped quantization noise to achieve coding gain.

MPEG-4 High-Efficiency AAC extends MPEG-4 AAC with the Spectral Band Replication tool. This new tool parametrically codes the high frequency components of the signal. Because the acuity of human perception of noise differs across the frequency band, the high-band parametric representation can introduce much larger errors, in a SNR sense, than perceptually shaped quantization noise and hence can achieve even greater coding gain.

The MPEG-D standards MPEG Surround and Spatial Audio Object Coding both incorporate parametric coding of the sound stage. MPEG Surround codes multiple audio channels (e.g. 5.1 channel audio programs) and Spatial Audio Object Coding codes multiple audio objects (e.g. lead guitar, bass guitar, drums, vocals) using highly compact parameters associated with each audio channel or audio object.

S. Quackenbush (✉)
Audio Research Labs, 336 Park Ave, Suite 200, Scotch Plains, NJ 07076
e-mail: srq@audioresearchlabs.com

L. Chiariglione (ed.), *The MPEG Representation of Digital Media*,
DOI 10.1007/978-1-4419-6184-6_7,

7.2 MPEG-4 Audio Coding

MPEG-4 Audio is a very broad initiative that proposed not only to increase the compression efficiency of audio coding, but also to support a number of new coding mythologies and functionalities [1]. It supports

- synthetic speech and audio engines (below 2 kb/s)
- a parametric coding scheme for low bit rate speech coding (2–4 kb/s)
- an analysis-by-synthesis coding scheme for medium bit rates (6–16 kb/s)
- a spectral bandwidth extension scheme for higher bit rates (32–48 kb/s)
- a subband/transform-based coding scheme for highest bit rates (64–128 kb/s)
- lossless coding for perfect reconstruction of the original signal (750 kb/s)

Of all of these coding tools and functionalities, two coding profiles have received widespread industry adoption: MPEG-4 AAC profile and MPEG-4 High Efficiency AAC profile. Each of these technologies will be described.

7.2.1 AAC

MPEG-4 AAC [1] builds upon MPEG-2 AAC [2], and in fact the basic coded representations are identical except that MPEG-4 AAC incorporates one additional tool: Perceptual Noise Substitution (PNS). A description of the PNS module follows:

Perceptual Noise Substitution (PNS) – this tool identifies segments of spectral coefficients that appear to be noise-like and codes them as random noise. It is extremely efficient in the that, for the segment, all that need be transmitted is a flag indicating that PNS is used and a value indicating the average power of the noise. The decoder reconstructs an estimate of the coefficients using a pseudo-random noise generator weighted by the signaled power value.

7.2.1.1 Subjective Quality

In 1998 MPEG conducted an extensive test of the quality of stereo signals coded by MPEG-2 AAC [3]. Since MPEG-4 AAC is identical to MPEG-2 AAC except for the inclusion of the PNS tool and since the MPEG-2 test was much more comprehensive than any test done for MPEG-4 AAC, it is referenced here. Critical test items were used and the listening test employed the BS-1116 Triple-Stimulus Hidden Reference (Ref/A/B) methodology [4], a methodology appropriate for assessment of very small impairments as would be found in nearly transparent coding technology. Thirty subjects participated in the test. The test concluded that for coding the stereo test items:

- AAC at 128 kb/s had performance that provided "indistinguishable quality" in the EBU sense of the phrase.

- AAC at 128 kb/s had performance that was significantly better than MPEG-1 Layer III at 128 kb/s.
- AAC at 96 kb/s had performance comparable to that of MPEG-1 Layer III at 128 kb/s.

7.2.2 *High Efficiency AAC*

The MPEG-4 High Efficiency AAC (HE-AAC) audio compression algorithm consists of an MPEG-4 AAC *core coder* augmented with the MPEG-4 Spectral Band Replication (SBR) tool [1, 5]. The encoder SBR tool is a pre-processor for the core encoder, and the decoder SBR tool is a post-processor for the core decoder, as shown in Fig. 7.1. The SBR tool essentially converts a signal at a given sampling rate and bandwidth into a signal at half the sampling rate and bandwidth, passes the low-bandwidth signal to the core codec, and codes the high-bandwidth signal using a compact parametric representation. The lowband signal is coded by the core coder, and the highband parametric data are inserted into the core coder bitstream and transmitted over the channel. The core decoder reconstructs the lowband signal and the SBR decoder uses the parametric data to reconstruct the highband data, thus recovering the full-bandwidth signal. This combination provides a significant improvement in compression efficiency relative to that of a full-band AAC codec.

7.2.2.1 SBR Principle

A perceptual audio coder, such as MPEG-4 AAC, provides coding gain by shaping the quantization noise such that it is always below the masking threshold. However, if the bit rate is not sufficiently high, the masking threshold will be violated, permitting coding artifacts to become audible. The usual solution adopted by perceptual coders in this case is to reduce the bandwidth of the coded signal, thus effectively increasing the available bits per sample to be coded. The result will sound cleaner but also duller due to the absence of high-frequency components.

The SBR tool gives perceptual coders an additional coding strategy (other than bandwidth reduction) when faced with severe bit rate restrictions. It exploits the human auditory system's reduced acuity to high-frequency spectral detail to permit it to parametrically code the high frequency region of the signal. When using the SBR tool, the lower frequency components of the signal (typically from 0 to between 5 and 13 kHz) are coded using the core codec. Since the signal bandwidth is reduced, the core coder will be able to code this signal without violating the masking threshold. The high frequency components of the signal are reconstructed as a transposition of the low frequency components followed by an adjustment of the spectral envelope. In this way, a significant bit rate reduction is achieved while maintaining comparable audio quality.

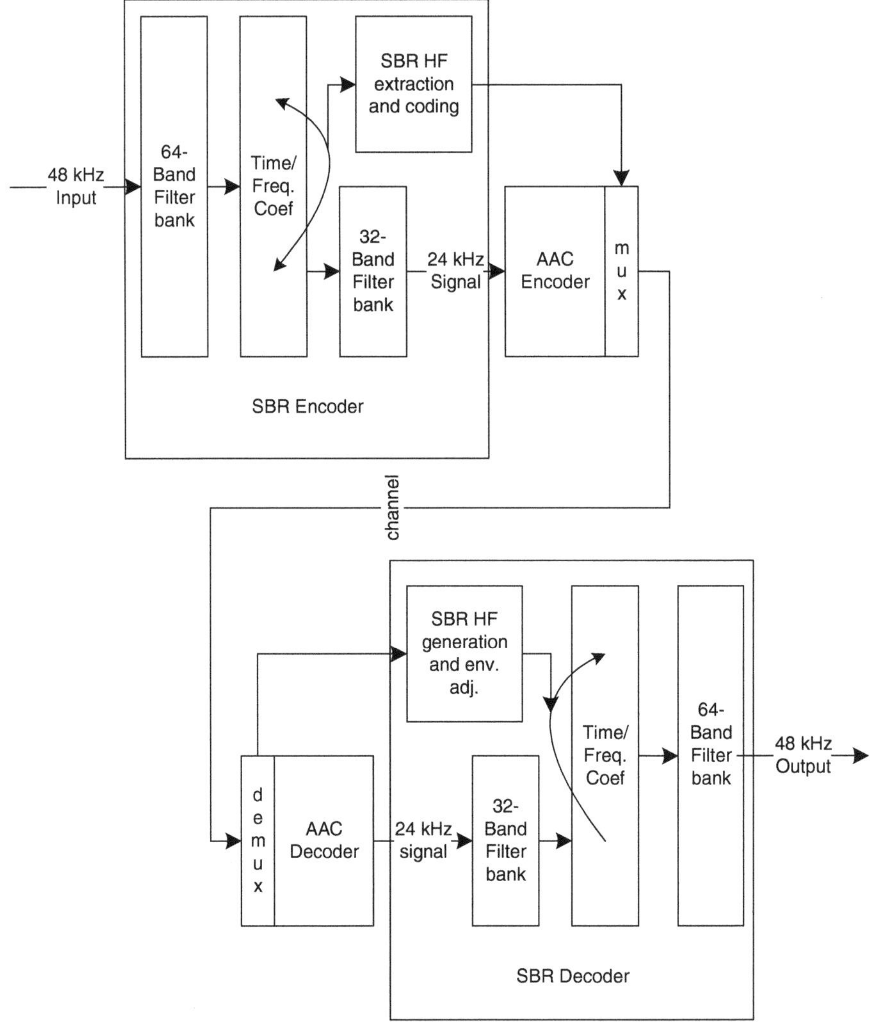

Fig. 7.1 Block diagram of HE-AAC encoder and decoder

7.2.2.2 SBR Technology

Using SBR, the missing high frequency region of a low pass filtered signal can be recovered based on the existing low pass signal and a small amount of side information, or control data. The required control data is estimated in the encoder based on the original wide-band signal. The combination of SBR with a core coder (in this case MPEG-4 AAC) is a dual rate system, where the underlying AAC encoder/decoder is operated at half the sampling rate of the SBR encoder/decoder. A block diagram of the HE-AAC compression system, consisting of SBR encoder and its sub-modules, AAC core encoder/decoder and SBR decoder and its sub-modules is shown in Fig. 7.1.

A major module in the SBR tool is a 64-band pseudo-quadrature mirror analysis/synthesis filterbank (QMF). Each block of 2048 PCM input samples processed by the analysis filterbank results in 32 subband samples in each of 64 equal-width subbands. The SBR encoder contains a 32-band synthesis filterbank whose inputs are the lower 32 bands of the 64 subbands and whose output is simply a band limited (to one-half the input bandwidth) and half-sampling rate version of the input signal. Actual implementations may use more efficient means to accomplish this, but the illustrated means provides a clearer framework for understanding how SBR works.

The key aspect of the SBR technology is that the SBR encoder searches for the best match between the signal in the lower subbands and those in the higher subbands (indicated by the curved arrow in the Time/Freq. Coef. box in the figure), such that the high subbands can be reconstructed by transposing the low subband signals up to the high subbands. This transposition mapping is coded as SBR control data and sent over the channel. Additional control parameters are estimated in order to ensure that the high frequency reconstruction results in a highband that is as perceptually similar as possible to the original highband. The majority of the control data is used for a spectral envelope representation. The spectral envelope information has varying time and frequency resolution such that it can control the SBR process in a perceptually relevant manner while using as small a side information rate as possible. Additionally, information on whether additional components such as noise and sinusoids are needed as part of the high-band reconstruction is coded as side information. This side information is multiplexed into the AAC bit stream (in a backward-compatible way).

In the HE-AAC decoder, the bit stream is de-multiplexed, the SBR side information is routed to the SBR decoder and the AAC information is decoded by the AAC decoder to obtain the half-sampling rate signal. This signal is filtered by a 32-band QMF analysis filterbank to obtain the low 32-subbands of the desired time/frequency coefficients. The SBR decoder then decodes the SBR side information and maps the low-band signal up to the high subbands, adjusts its envelope and adds additional noise and sinusoids if needed. The final step of the decoder is to reconstruct the output block of 2,048 time-domain samples using a 64-band QMF synthesis filterbank whose inputs are the 32 low subbands resulting from processing the AAC decoder output and 32 high subbands resulting from the SBR reconstruction. This results in an up-sampling by a factor of two.

7.2.2.3 Subjective Quality

MPEG conducted several tests on the quality of HE-AAC, including a final verification test [6]. There were two test sites in the test using stereo items: France Telecom and T-Systems Nova, together comprising more than 25 listeners, and four test sites in the test using mono items: Coding Technologies, Matsushita, NEC, and Panasonic Singapore Laboratories, together comprising 20 listeners. In all cases the MUSHRA test methodology was used [7], which is appropriate for assessment of intermediate quality coding technology.

The overall conclusions of the test were that, for the stereo items:

- The audio quality of High Efficiency AAC at 48 kbps is in the lower range of "Excellent" on the MUSHRA scale.
- The audio quality of High Efficiency AAC at 48 kbps is equal or better than that of MPEG-4 AAC at 60 kbps.
- The audio quality of High Efficiency AAC at 32 kbps is equal or better than that of MPEG-4 AAC at 48 kbps.

7.3 MPEG-D Audio Coding

MPEG-4 provided standards for audio coding with a rich set of functionalities. MPEG-D is more in the tradition of MPEG-1 and MPEG-2, in which compression is the primary goal. The first standard in MPEG-D is MPEG Surround, which is a technology for the compression of multi-channel audio signals [8, 9]. It exploits human perception of the auditory sound stage, and in particular preserves and codes only those components of a multi-channel signal that the human auditory system requires so that the decoded signal is perceived to have the same sound stage as the original signal [10].

7.3.1 MPEG Surround

The MPEG Surround algorithm achieves compression by encoding an N-channel (e.g., 5.1 channel) audio signal as a stereo (or mono) audio signal plus side information in the form of a secondary bit stream which contains "steering" information. The stereo (or mono) signal is encoded and the compressed representation sent over the transmission channel along with the side information, such that a spatial audio decoder can synthesize a high quality multi-channel audio output signal in the receiver. Although the audio compression technology is quite flexible, such that it can take N input audio channels and compress those to one or two transmitted audio channels plus side information, discussed here is only a single example configuration, which is the case of coding a 5.1 channel signal as a compressed stereo signal plus some side information. This configuration is illustrated in Fig. 7.2.

Referring to the figure, the MPEG Surround encoder receives a multi-channel audio signal, x_1 to $x_{5.1}$, where x is a 5.1 channel (left, center, right, left surround, right surround, LFE) audio signal. Critical in the encoding process is that a "downmix" signal is derived from the multi-channel input signal, and that this downmix signal is compressed and sent over the transmission channel rather than the multi-channel signal itself. A well-optimized encoder is able to create a downmix that

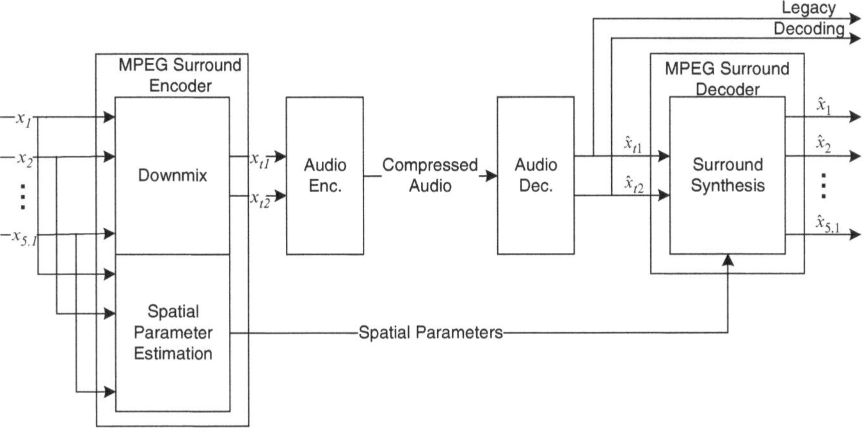

Fig. 7.2 Principle of MPEG surround

is, by itself, a faithful stereo equivalent of the multi-channel signal, and which also permits the MPEG Surround decoder to create a perceptual equivalent of the multi-channel original. The downmix signal x_{t1} and x_{t2} (2-channel stereo) is compressed and sent over the transmission channel. The MPEG Surround encoding process is agnostic to the core audio compression algorithm used. It could be any of a number of high-performance compression algorithms such as MPEG-4 High Efficiency AAC, MPEG-4 AAC, MPEG-1 Layer III or even PCM. The audio decoder reconstructs the downmix as $\hat{x}_{t1}$ and $\hat{x}_{t2}$. Legacy systems would stop at this point and present the decoded downmix, otherwise this decoded signal plus the spatial parameter side information are sent to the MPEG Surround decoder which reconstructs the multi-channel signal $\hat{x}_1$ to $\hat{x}_{5.1}$.

The heart of the encoding process is the extraction of spatial cues from the multi-channel input signal that capture the salient perceptual aspects of the multi-channel sound image (this is also referred to as "steering information" in that it indicates how to "steer" sounds to the various loudspeakers). Since the input to the MPEG Surround encoder is a 5.1-channel audio signal that is mixed for presentation via loudspeaker, sounds located at arbitrary points in the sound stage are perceived as phantom sound sources located between loudspeaker positions. Because of this, MPEG Surround computes parameters relating to the differences between each of the input audio channels. These cues are comprised of:

- Channel Level Differences (CLD), representing the level differences between pairs of audio signals.
- Inter-Channel Correlations (ICC), representing the coherence between pairs of audio signals.
- Channel Prediction Coefficients (CPC), representing the ability to predict an audio signal from others.
- Prediction error (or residual) signals, representing the error in the parametric modeling process relative to the original waveform.

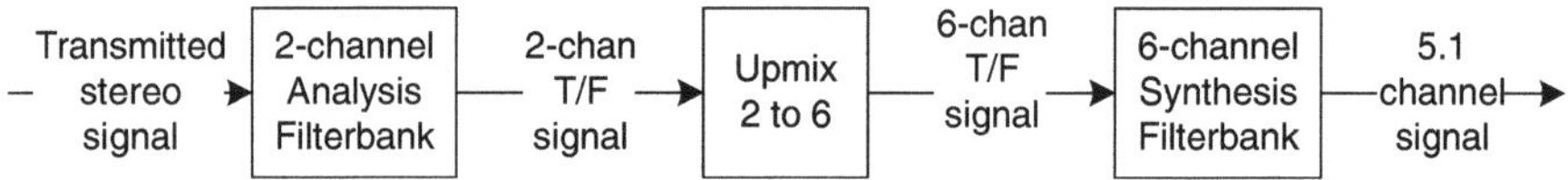

Fig. 7.3 Block diagram of MPEG surround spatial synthesis

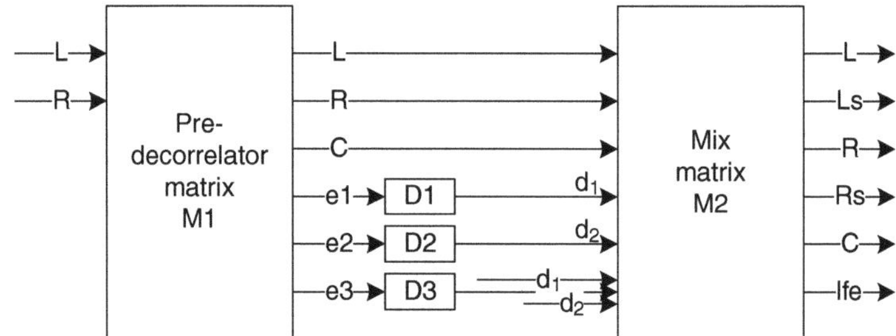

Fig. 7.4 Block diagram of MPEG surround upmix

A key feature of the MPEG Surround technique is that the transmitted downmix is an excellent stereo presentation of the multi-channel signal. Hence legacy stereo decoders do not produce a "compromised" version of the signal relative to MPEG Surround decoders, but rather the very best version that stereo can render. This is vital, since stereo presentation will remain pervasive due to the number of applications for which listening is primarily via headphones (e.g. portable music players).

In the decoder, spatial cues are used to upmix the stereo transmitted signal to a 5.1 channel signal. This operation is done in the time/frequency (T/F) domain as is shown in Fig. 7.3. Here an analysis filterbank converts the input signal into two channels of T/F representation, where the upmix occurs (schematically illustrated in Fig. 7.4), after which the synthesis filterbank converts the six channels of T/F data into a 5.1 channel audio signal. The filterbank must have frequency resolution comparable to that of the human auditory system, and must be oversampled so that processing in the time/frequency domain does not introduce aliasing distortion. MPEG Surround uses the same 64-band filterbank as is used in HE-AAC, but with division of the lowest frequency bands into additional subbands using an MDCT.

The upmix process applies mixing and decorrelation operations to regions of the stereo T/F signal to form the appropriate regions of the 5.1 channel T/F signals. This can be modeled as two matrix operations (M1 and M2) plus a set of decorrelation filters (D_1–D_3), all of which are time-varying. Note that in the figure the audio signals (L, R, C, Ls, Rs, lfe) are in the T/F domain.

In addition to encoding and decoding multi-channel material via the use of side information, as just described, the MPEG Surround standard also includes operating modes that are similar to conventional matrix surround systems. It can encode a multi-channel signal to a matrixed stereo signal, which it can decode back to the multi-channel signal. This mode is MPEG Surround with "zero side information," and is fully interoperable with conventional matrixed surround systems. However,

MPEG Surround has an additional feature in this mode: it can produce a matrix-compatible stereo signal and, if there is a side information channel, it can transmit information permitting the MPEG Surround decoder to "un-do" the matrixing, apply the normal MPEG Surround multi-channel decoding, and present a multi-channel output that is superior to that of a matrix decoder upmix.

MPEG Surround has a unique architecture that, by its nature, can be a bridge between the distribution of stereo material and the distribution of multi-channel material. The vast majority of audio decoding and playback systems are stereo, and MPEG Surround maintains compatibility with that legacy equipment. Furthermore, since MPEG Surround transmits an encoded stereo (or mono) signal plus a small amount of side information, it is compatible with most transmission channels that are currently designed to carry compressed stereo (or mono) signals. For multi-channel applications requiring the lowest possible bit rate, MPEG Surround based on a single transmitted channel can result in a bit rate saving of more than 80% as compared to a discrete 5.1 multi-channel transmission.

7.3.1.1 Subjective Quality

In 2007 MPEG conducted a test of the quality of MPEG Surround [11]. It was conducted at several test sites and comprised in total more than 40 listeners. Critical five-channel test items were used that demonstrated an active sound stage and the MUSHRA test methodology was employed.

The test report showed that, when coding five-channel signals:

- At 160 kb/s, MPEG Surround in combination with the HE-AAC downmix coder achieved performance that was in the middle of the "Excellent" quality range on the MUSHA scale.
- At 64 kb/s, MPEG Surround in combination with the HE-AAC downmix coder achieved performance that was at the high end of the "Good" quality range on MUSHA scale.

In addition, evaluations conducted during the standardization process showed that MPEG Surround, when employing the residual coding tool, has the ability to scale from the high range of "Good" quality to the high range of "Excellent" quality as its bitrate increases from 160 to 320 kb/s. In particular:

- At 320 kb/s, MPEG Surround in combination with the AAC downmix coder achieves performance that is comparable to that of AAC when coding the five discrete channels at 320 kb/s.

7.3.2 Spatial Audio Object Coding

Spatial Audio Object Coding (SAOC) is a technique for efficient coding and flexible rendering of multiple audio objects within a sound stage [12, 13]. While MPEG

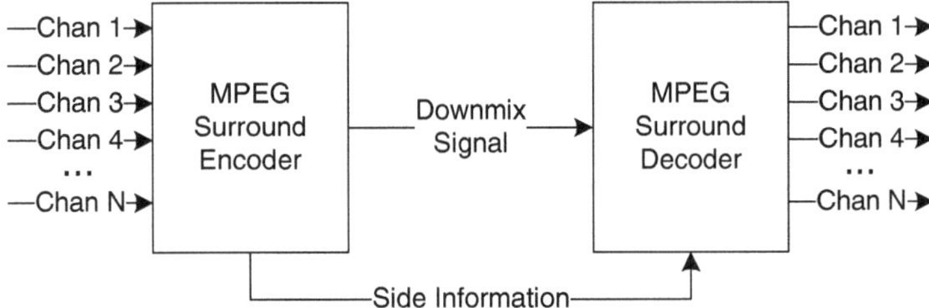

Fig. 7.5 Channel-based MPEG surround

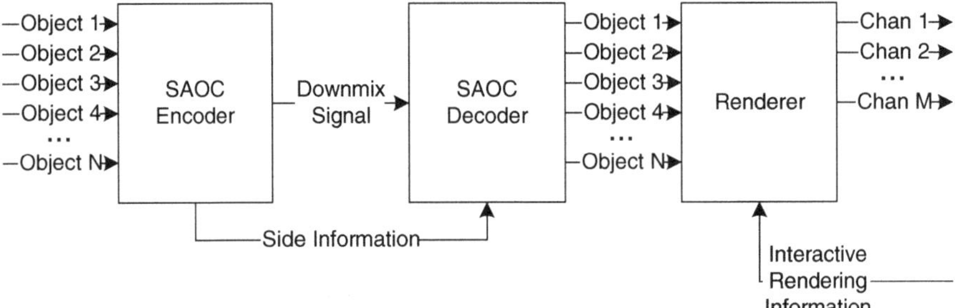

Fig. 7.6 Object-based SAOC

Surround encodes and reconstructs multiple audio channels, as shown in Fig. 7.5, SAOC encodes multiple audio *objects*, reconstructs those objects and then renders them, possibly according to the user's preferences, as shown in Fig. 7.6.

In SAOC the rendering operation is simply a matrix multiplication that takes N objects and produces M channels as $Y = AS$, where Y is a block of the M output channel signals, A is the rendering matrix and S is a block of the N object signals. Rendering matrix A can be transmitted or obtained as user information and in each case can be updated dynamically.

Conceptually, the SAOC decoder can be thought of as a pre-processor to an MPEG Surround decoder, and in fact is specified as such for multi-channel outputs. For stereo outputs (as in personal music re-mixing), additional MPEG Surround processing is not required. Both SAOC and MPEG Surround insert side-information into a compressed downmix signal. In the case of SAOC, the SAOC pre-processor extracts the object-based side information from the bitstream of the compressed downmix, uses the rendering information to convert it to channel-based side information, re-inserts it into the bitstream and passes it to the MPEG Surround decoder. This is illustrated in Fig. 7.7.

SAOC is a parametric multiple audio object coding technique. It can transmit N audio objects as mono or stereo downmix signal plus side information. For example rather than encoding four objects (lead guitar, bass guitar, drums, vocals) as four

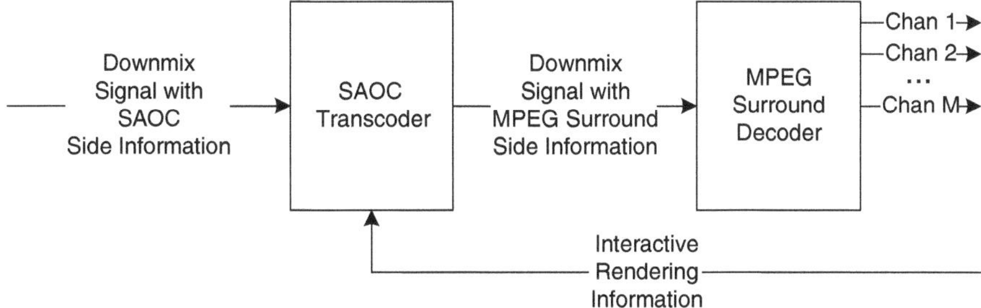

Fig. 7.7 SAOC Decoder as a tandem of SAOC Transcoder and MPEG surround decoder

Table 7.1 Frequency resolution of SAOC filterbank

Band	Frequency range (Hz)	Bandwidth (Hz)
0	0–86	86
1	86–172	86
2	172–258	86
3	258–345	86
4	345–517	172
5	517–689	172
6	689–861	172
7	861–1,034	172
8–68	1,034–22,050	345

Figures based on 44.1 kHz sampling rate

monophonic signals, SAOC can be used to encode them as a stereo downmix signal plus a relatively low bitrate of side information, thus providing considerable compression gain.

In the SAOC bitstream, the audio objects are typically represented by the following parameters (although additional parameters may apply in special situations):

- Object Level Difference (OLD), which describes the energy of one object relative to the object with the greatest energy in a given time and frequency band.
- Inter-Object Cross-correlation (IOC), which describes the similarity, or cross-correlation, for two objects in a given time and frequency band.

These parameters are defined in the complex time/frequency (T/F) domain, which results when filtering the input signal by a 64-band complex-valued Quadrature Mirror Filter (QMF) bank. Processing of a block of input signal typically produces a 64 (frequency bands) by 32 (time samples per frequency band) array of T/F information. In order to obtain a resolution at low frequencies that is comparable to that of resolution of the human auditory system, the first 8 bands are further split giving a frequency resolution as show in Table 7.1. This is the same filterbank as is used in MPEG Surround, thus permitting computationally efficient tandeming of SAOC and MPEG Surround.

Parameters are specified over *tiles* or contiguous sub-blocks of the T/F array, and the size of these tiles are dynamically scalable, permitting the side information data rate for the coded parameters to be easily varied. All parameters are quantized and then entropy coded prior to insertion into the downmix signal bitstream. The total SAOC side information data rate depends on the number of audio objects represented, but typical side information data rate is approximately 3 kb/s per audio object.

From the SAOC parameters (OLD, IOC and perhaps others) one can derive an approximation of the object covariance, $E = SS^*$ where S is a block of the object signals, * is the conjugate operator (since the object signals are complex) and the elements of E are the object powers and cross-correlations [14]. The target output channel covariance, R, can be expressed as

$$R = YY^* = AS(AS)^* = ASS^*A = AEA^*$$

where Y is the audio channel output signal and A is the rendering matrix for that channel.

For brevity only the SAOC transcoder for a mono downmix will be described here. The MPEG Surround decoder employs a tree-structured parameterization based on a cascade of One-To-Two (OTT) upmix modules as described in [8]. This tree-structure leads to an upmix of the mono downmix into e.g. a 5.1 channel presentation. Associated with each OTT module are the MPEG Surround Channel Level Differences (CLD) and Inter-Channel Correlations (ICC). Estimation of CLDs and ICCs from the SAOC parameters OLD and IOC is performed separately for each OTT module by deriving corresponding rendering sub-matrices from the full rendering matrix. Extracting the CLD and ICC parameters for the mono downmix case is straight forward: the diagonal elements of R are the object powers which are then easily translated into the objects' absolute power or Arbitrary Downmix Gain (ADG) parameter and channel level differences (CLD) parameter; the off-diagonal elements of R, r_{ij} translate directly into the ICC parameters for channel pair i and j.

7.3.2.1 Teleconferencing Application

In a typical telephone conversation there is one remote party and if that person is speaking, you know who it is. However, in a typical teleconference, there are many remote parties, typically at several different locations. This may lead to considerable difficulty in determining who is speaking. A teleconference using SAOC technology provides a remedy to this difficult situation.

Imagine that there are four conference participants, as shown in Fig. 7.8. Although not mentioned in this brief overview, SAOC permits a very efficient bridging in the compressed domain, such that if each conference participant sends a SAOC bitstream to the conference bridge (labeled "SAOC System" in the figure), they can be combined into a single SAOC bitstream with all objects (i.e. conference participants) in the compressed downmix signal and all object parameters in the side information.

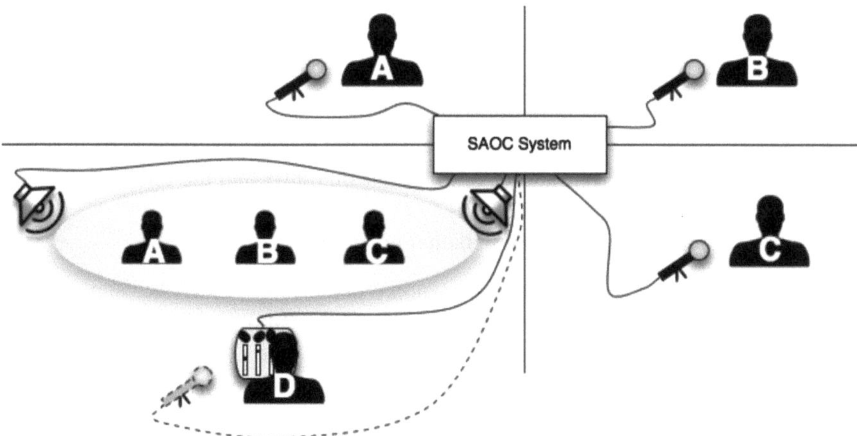

Fig. 7.8 Scenario for a SAOC teleconference

This combined SAOC bitstream is then sent from the conference bridge to each participant site. Note that the bridge receives a compressed downmix (typically mono) signal and sends a compressed downmix (mono) signal. In each case the compressed signal contains a small data rate of SAOC side information, so that the bandwidth of a SAOC teleconference is no different from that of a conventional teleconference.

However, at the SAOC decoder each conference participant (e.g. participant "D" in the figure), can interactively specify a rendering matrix such that the SAOC decoder will distribute the remote participants in a stereo or multi-channel sound stage. In this way, the sound stage cues (e.g. left, centre, right) can be used to much more easily identify the participant that is currently talking. This lowers "teleconferencing fatigue" and permits more of the participant's energy to be dedicated to the substance of the conference.

7.3.2.2 Personal Re-Mix Application

The SAOC bitstream contains a downmix signal and parameters associated with each object in that downmix. If the compressed downmix signal in SAOC is mono or stereo, it is then backwards-compatible with any music player that can decode the compressed audio format (e.g. MP3 or AAC music players). The SAOC bitstream typically contains a default rending matrix, which in the case of a music signal, represents the production music mix. However, an SAOC decoder can receive rendering information from the user so that he or she can interactively re-mix the music. For example, he or she can make a "dance mix" (e.g. boost the percussion) or a "relaxing mix" (less percussion, more vocals). Alternatively, the user could alter the sound stage: move the guitar from right to left and move vocals from centre to right. All this is made possible by SAOC's object-based signal representation.

7.3.2.3 Subjective Quality

In 2010 MPEG conducted a test of the quality of SAOC systems configured for personal music remix applications and for teleconferencing applications [15]. It was conducted at several test sites and comprised in total more than 30 listeners. Critical test items were used appropriate for both the music remix and the teleconferencing application scenarios.

The test report showed that in a personal music remix scenario:

- SAOC technology with a stereo downmix coded with HE-AAC significantly outperforms (by about 20 points on the MUSHRA scale) discrete channel coding using HE-AAC when both operate at the same data rate of 64 kbit/s, achieving a subjective quality in the upper region of the "Good" range on the MUSHRA scale.

The test report showed that in a teleconferencing scenario:

- The SAOC technology with a mono downmix coded with AAC-ELD delivers audio quality statistically equivalent to the quality provided by legacy technology (i.e. separately encoded streams), but does so using less than half the bitrate (i.e. 40 kbit/s for SAOC vs. 94 kbit/s for discrete channel coding), achieving a subjective quality in the "Good" range on the MUSHRA scale.
- The performance of SAOC is similar for stereo and multi-channel reproduction.

Acknowledgements The author would like to thank Oliver Hellmuth, Fraunhofer Institute for Integrated Circuits, for permission to use a number of his SAOC-related figures.

References

1. ISO/IEC, "Information technology – Coding of audio-visual objects – Part 3: Audio," ISO/IEC Int. Std. 14496–3:2009. (Available at http://www.iso.org/iso/store.htm).
2. ISO/IEC, "Information technology – Generic coding of moving pictures and associated audio information – Part 7: Advanced Audio Coding (AAC)," ISO/IEC Int. Std. 13818–7:2006. (Available at http://www.iso.org/iso/store.htm).
3. Meares, D, Watanabe, K and Scheirer, E, Report on the MPEG-2 AAC Stereo Verification Tests, ISO/IEC JTC1/SC29/WG11/N2006. February 1998. (Available at http://mpeg.chiariglione.org/quality_tests.htm).
4. ITU-R Recommendation BS. 1116–1, "Methods for the subjective assessment of small impairments in audio systems including multichannel sound systems," Geneva, Switzerland, 1997.
5. Dietz, M., Liljeryd, L., Kjörling, K., Kunz, O., "Spectral Band Replication, a novel approach in audio coding," 112th AES Convention Proceedings, Audio Engineering Society, May 10–13, 2002, Munich, Germany. Preprint 5553.
6. ISO/IEC JTC1/SC29/WG11/N6009, Report on the Verification Tests of MPEG-4 High Efficiency AAC. (Available at http://mpeg.chiariglione.org/quality_tests.htm).
7. ITU-R Recommendation BS 1534-1, "Method for the subjective assessment of intermediate quality level of coding systems: MUlti-Stimulus test with Hidden Reference and Anchor (MUSHRA)," Geneva, Switzerland, 1998–2000 (Available at http://ecs.itu.ch).
8. J. Herre, et al., "MPEG Surround – The ISO/MPEG Standard for Efficient and Compatible Multi-Channel Audio Coding," in AES 122nd Convention, Vienna, Austria, May 2007, preprint 7084.

9. ISO/IEC, "Information technology – MPEG audio technologies – Part 1: MPEG Surround," ISO/IEC Int. Std. 23001–1:2007. (Available at http://www.iso.org/iso/store.htm).
10. J. Breebaart and C. Faller, *Spatial audio processing: MPEG Surround and other applications*. John Wiley & Sons, Chichester, 2007.
11. ISO/IEC JTC1/SC29/WG11/N8851, Report on MPEG Surround Verification Tests (Available at http://mpeg.chiariglione.org/quality_tests.htm).
12. O. Hellmuth, et al. "MPEG Spatial Audio Object Coding – The ISO/MPEG Standard for Efficient Coding of Interactive Audio Scenes," 129th AES Convention, November 2010, San Francisco, preprint 8264.
13. ISO/IEC, "Information technology – MPEG audio technologies – Part 2: Spatial Audio Object Coding (SAOC)," ISO/IEC Int. Std. 23002–1:2010. (Available at http://www.iso.org/iso/store.htm).
14. J. Engdegard, et al. "Spatial Audio Object Coding (SAOC) – The Upcoming MPEG Standard on Parametric Object Based Audio Coding," 124th AES Convention, May 2008, Amsterdam, preprint 7377.
15. ISO/IEC JTC1/SC29/WG11/N11657, Report on Spatial Audio Object Coding Verification Tests (Available at http://mpeg.chiariglione.org/quality_tests.htm).

Chapter 8
MPEG Audio Compression Future

Schuyler Quackenbush

8.1 Introduction

We have seen that MPEG-1 and MPEG-2 Audio coders used perceptually shaped quantization noise as the primary tool for achieving compression. MPEG-4 High-Efficiency AAC [1] added parametric coding of the upper spectrum region with the Spectral Band Replication tool and parametric coding of two-channel signals with the Parametric Stereo tool. The MPEG-D standards MPEG Surround [2] and Spatial Audio Object Coding [3] incorporate parametric coding of the sound stage using level, time and coherence parameters for regions of the time/frequency signal. The common thread in all of these tools is that they model and exploit how humans perceive sound.

MPEG-D Unified Speech and Audio Coding [4–6] incorporates all of these models of sound perception, and additionally incorporates a model of sound production, specifically that of human speech.

The MPEG Audio subgroup has a rich history of accomplishments in creating music coding technology, as discussed in earlier chapters of this book. At higher bit-rates at which MPEG standards are rated as transparent (e.g., 128 kbps for a stereo signal using AAC), MPEG technology can represent arbitrary sounds, including the human voice, with excellent quality. However, as higher compression becomes more and more important for delivering services over relatively expensive wireless channels, MPEG has developed standards that are able to deliver audio signals at lower and lower bit-rates (e.g., 32 kbps for a stereo signal using HE-AAC V2 [1]).

The very successful MPEG family of AAC specifications are all built upon the transform-based perceptual coder. The transform tool works well with a variety of signals, in that signals with little redundancy have considerable irrelevancy.

S. Quackenbush (✉)
Audio Research Labs, 336 Park Ave, Suite 200,
Scotch Plains, NJ 07076
e-mail: srq@audioresearchlabs.com

L. Chiariglione (ed.), *The MPEG Representation of Digital Media*,
DOI 10.1007/978-1-4419-6184-6_8,

For example, non-stationary, noise-like signals have almost no redundancy across the transform block and hence no transform coding gain but require only 5–6 dB of coding SNR per critical band. Bits can be allocated relatively uniformly because the required coding SNR is low. Conversely, signals with considerable redundancy have little irrelevancy. For example, stationary tone-like signals require 20–35 dB of coding SNR per critical band. The large transform gain for these signals can be used to selectively allocate bits where the greatest coding SNR is required.

The transform tool is not uniformly the best coding tool at higher levels of compression, especially for the speech signal. Speech is highly non-stationary, and segments of the speech signal can have very different statistics:

- Un-voiced speech (e.g., fricatives) can be a good match to a transform coder in that these regions are noise-like.
- Voiced speech might appear to be a good match to a transform coder in that its spectrum is approximately stationary. However, closer inspection reveals that the speech pitch fundamental frequency (and its associated harmonics) varies considerably over the typical transform block time. This variation can be a pitch frequency modulation (e.g., vibrato) or a pitch frequency glide (e.g., changes in intonation).
- Plosives or stops inherently have low transform gain in that they are short-time events (e.g., silence or onset) with statistics that change significantly within a single transform block.

Often a transform tool does not have high performance when coding speech because it is configured for high frequency resolution and therefore has inadequate time resolution. In order to achieve high transform coding gain for stationary tonal signals, the transform needs a frequency resolution that is comparable to that of the narrowest critical band (approximately 100 Hz at low frequencies). This fact dictates a time resolution of 20–40 ms, but speech is rarely stationary over this period of time. The result is very audible quantization noise.

8.2 Unified Speech and Audio Coding

8.2.1 Genesis

The previous section discussed reasons why a typically configured transform-based perceptual coder is unable to deliver high coding performance for signals such as speech. However, there is a very rich history of speech coding technology and many examples of technology that does deliver high performance when coding speech signals. Even with this history, MPEG audio experts observed that there was no single, unified coding architecture that could deliver uniformly high performance for both speech and music signals. MPEG decided to undertake a project to determine if such a unified coding architecture was feasible.

In October 2007 WG11 issued a Call for Proposal on Unified Speech and Audio Coding [7]. The Call required that the performance of submitted technology be evaluated when coding test items representative of the three content categories – speech; music; and mixed speech and music – at operating points ranging from 12 kbps for mono signals to 64 kbps for stereo signals. Most importantly, it required that the final performance of the MPEG specification be measured relative to two state-of-the-art codecs. Specifically, that the performance of the MPEG specification be comparable to the performance of the better of 3GPP AMR-WB+ [8] or MPEG-4 High-Efficiency Advanced Audio Coding v2 [1] at every operating point for each test item. Using this criterion, the Call envisioned that proposals would have both consistent and state-of-the-art compression performance for any mix of speech and music content.

WG11 evaluated responses to the Call in July 2008 and found that the best proposed technology, which was selected as the foundation on which the Unified Speech and Audio Coding (USAC) standard was built, already satisfied the final performance criterion for the work item. This situation was very promising, and the subsequent collaborative phase of development further improved the technology, which became an international standard in late 2011.

8.2.2 Architecture

8.2.2.1 AAC Foundation

The Unified Speech and Audio Coding (USAC) technology was built upon the MPEG-4 Advanced Audio Coding (AAC) [1, 9] and High Efficiency AAC (HE-AAC) codecs [1, 10, 11]. In order to achieve the highest level of performance, it was decided that backwards compatibility to AAC or HE-AAC was not a requirement. Hence, the new standardization activity took the opportunity to review the performance of all tools and revise them if greater compression performance could be obtained.

In order to understand the USAC architecture, it is instructive to review the AAC architecture. AAC is a classic perceptual coder for which a simplified block diagram is shown in Fig. 8.1. In the AAC encoder, overlapping blocks of input signal are processed by a modified discrete cosine transform (MDCT). The transform coefficients are scaled by scalefactors that are determined by the masking threshold estimated in the perceptual model; the scaled coefficients are quantized and then entropy coded. In USAC this is referred to as frequency domain (FD) coding.

8.2.2.2 Transform Coded Excitation Tools

As discussed previously, the AAC architecture is not always able to effectively code the speech signal. One tool that USAC added to AAC was linear-prediction

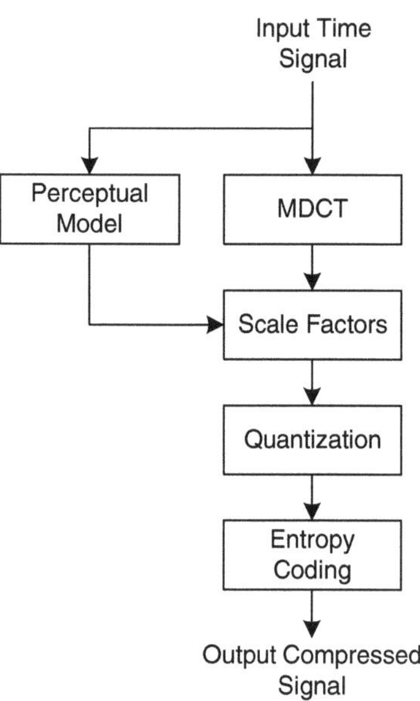

Fig. 8.1 AAC encoder block diagram

transform-coded-excitation (TCX) [12], similar to what is used in the AMR-WB+ coding specification [8, 13]. TCX incorporates short-term prediction to model how the human vocal tract shapes the speech signal. As such, it is an explicit source model as contrasted to the perception models (i.e. psychoacoustic models) that the MPEG AAC family of coders use.

The left panel (a) of Fig. 8.2 shows a classic TCX encoder architecture, while the right panel (b) shows a modified but equivalent architecture. Note that (a) is shown with an MDCT and entropy coder, while AMR-WB+ uses a fast Fourier transform (FFT) and lattice vector quantizer, but the principles of the TCX coder are the same. Details such as the perceptual model are not shown. In the TCX encoder of panel (a), a linear prediction coding (LPC) model is estimated from the input signal every block; the signal is inverse filtered using this model to produce a whitened signal which is transformed using an MDCT. The transform coefficients are quantized and entropy coded.

An equivalent architecture is shown in panel (b), in which the order of inverse filter and transform is interchanged. In this case, the LP whitening is achieved by scaling the transform coefficients by the LP envelope, or system function, which is simply the transform of the LP impulse response. In fact, the whitening is a simple scaling only if the LP model is constant across the interval of the transform block. In general, the whitening is implemented by a recursive filtering of the MDCT coefficients that reflects the fact that the LP model is changing over time.

The advantage of this modified TCX architecture is apparent in Fig. 8.3. The AAC and TCX coding architectures use many of the same tools, and Fig. 8.3 shows how

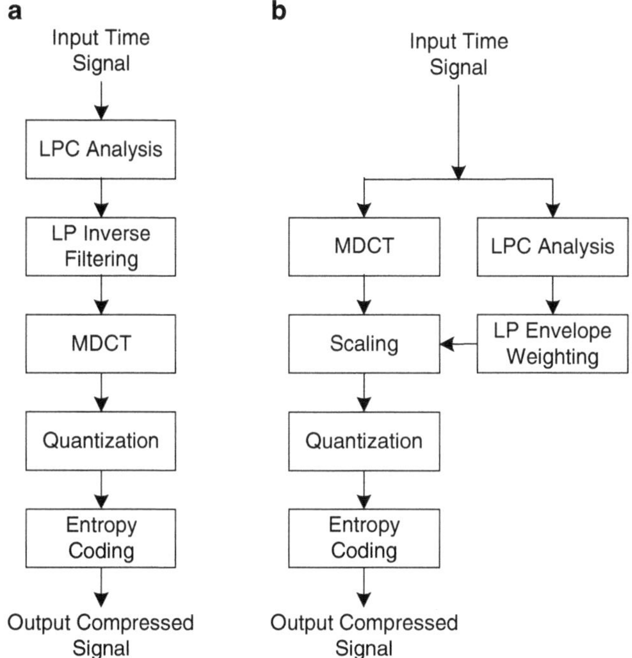

Fig. 8.2 Transform Coded Excitation coder. (**a**) Shows the classic design and (**b**) shows a modified but equivalent design

they can be merged together to eliminate duplication – the leftmost branch is the AAC scaling path and the rightmost branch is the TCX scaling path. Not shown in the figure is the decision logic required to adaptively select which coding branch is used for each input signal block.

Previous chapters have reviewed that the MDCT in AAC can adaptively switch frequency resolution by using one 2,048- or a sequence of eight 256-length windows to process a pair of adjacent blocks of 1,024 input samples (with 50% overlap of the transform windows) and produce 1,024 transform coefficients. Similarly, the MDCT windows used in the TCX coding mode have varying time/frequency resolution, and the window properties TCX are shown in Table 8.1. Note that some TCX windows have zero-padding at each end and therefore have a shorter "effective window length" than indicated by the number of transform coefficients produced (the window length is typically equal to twice the number of coefficients).

In the unified AAC/TCX architecture, the transition between coding modes (or different window lengths within a coding mode) is handled by the MDCT, where the overlap of the sloping portion of the windows inherently smooths the transition between modes and cancels the time-domain aliasing [14]. The transition between various windows and modes is shown in Fig. 8.4. In each panel the vertical axis is

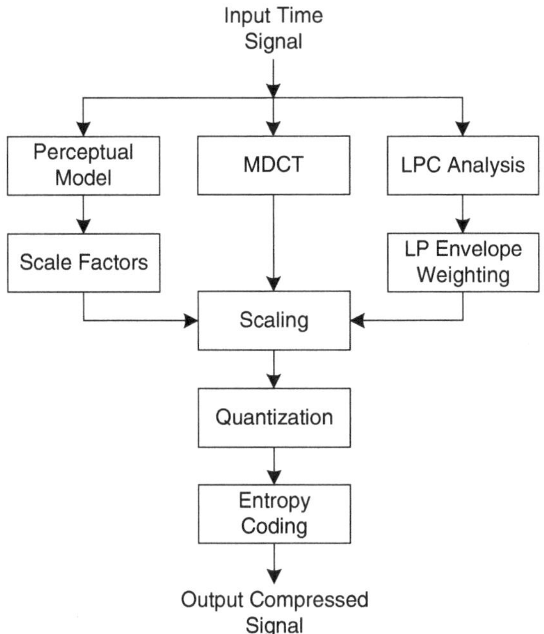

Fig. 8.3 Merging the AAC and modified TCX architectures

Table 8.1 MDCT properties in TCX coding mode

Name	Effective window length	Zero pad	Num. coef.
TCX 256	512	0	256
TCX 512	768	128	512
TCX 1024	1,280	384	1,024

amplitude and the horizontal axis is time, where each of the vertical dashed lines represents an interval of one coding block of samples.

The panel labelled FD shows the transition between long and short blocks in FD mode. In this case, the FD mode start window has a trailing slope of length 128 so it can cancel the time-aliasing with the leading short window. Plots labelled TCX 256, TCX 512 and TCX 1024 show the transition between long blocks in FD mode and the TCX windows. In this case, the FD mode start window has a trailing slope of length 256 so it can cancel the time-aliasing with the leading edge of the TCX windows, which all have a slope of length 256 and varying intervals of unity amplitude.

The various window lengths permit USAC to have an adaptive time/frequency resolution as is needed to code an arbitrary mix of speech and music content. Table 8.2 shows the effective window lengths (in samples and ms) and the corresponding frequency resolution for each window length (in Hz). Note that the frequency resolution of the zero-padded windows is less than that of the nominal window length (i.e., equal to twice the number of transform coefficients) would suggest, due to the increased main lobe width of the window and associated frequency-domain smearing.

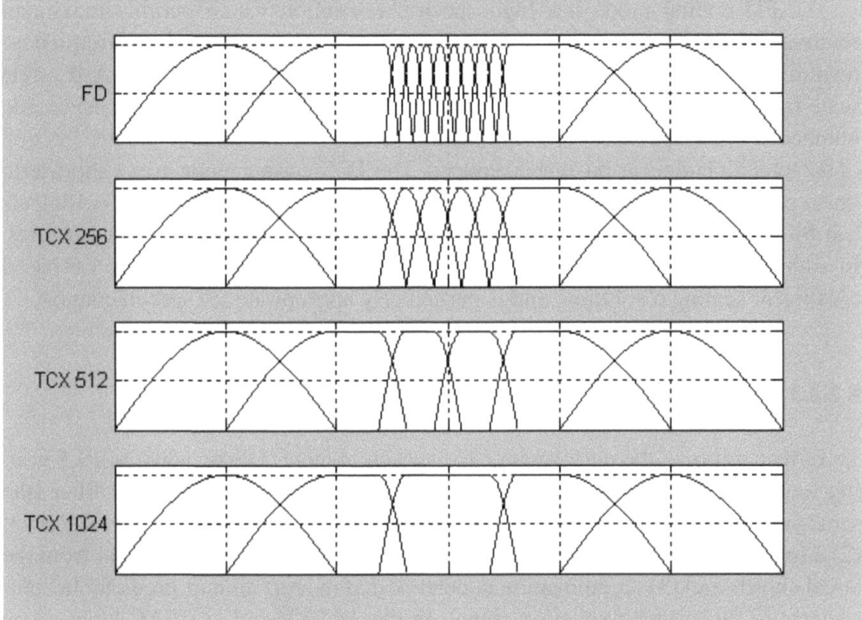

Fig. 8.4 Transitions between various window lengths and coding modes

Table 8.2 Effective time/frequency resolution of the FD and TCX coding modes

	Effective window length		Effective frequency resolution
Coding mode	Samples	ms	Hz
FD long	2,048	85.3	11.7
TCX 1024	1,280	53.3	37.5
TCX 512	768	32.0	62.5
TCX 256	512	21.3	93.8
FD short	256	5.3	187.5

This table assumes a core coder sampling rate of 24 kHz and a coding block length of 1,024

The table assumes a 24 kHz sampling rate and a USAC core coder block length of 1,024 samples. Note that USAC supports a wide range of sampling rates from 8 kHz to 96 kHz and 24 kHz is selected here only as an example. Additionally, USAC supports an operation mode where the block length is equal to 768 samples. In this case, the values in Table 8.1 are simply multiplied by a factor of ¾. The sampling rate of 24 kHz is used (rather than, e.g., 48 kHz) because at medium to low bit-rates where the TCX coding mode is often used, eSBR (see Fig. 8.7) is typically also used. In a typical configuration the eSBR module will drop the sampling rate presented to the core coder by a factor of 2.

The FD coding mode has high spectral resolution which permits maximum removal of signal redundancy. Such a configuration requires high-frequency-resolution scaling of the transform coefficients, as is the case with the AAC-style scale factors in USAC. However, coding the scale factors consumes a considerable number of bits (especially for short blocks), which become an ever greater fraction of the total bit budget as bit-rate is reduced. The TCX coding mode uses a short-term linear prediction model to scale the transform coefficients, and these LP coefficients can by very efficiently represented, which is needed at medium to low bit-rates. In addition, the TCX time/frequency resolution is well matched by the LP-based coefficient scaling resolution, and is particularly appropriate for speech content.

8.2.2.3 ACELP Coding Tools

To further improve the performance for speech content, USAC adds ACELP coding tools [15, 16]. These tools incorporate (1) a short-term prediction filter (the same as in the TCX modes) that models the shaping done by the human vocal tract; (2) a long-term prediction filter that models the periodic excitation signal from the vocal chords and (3) an innovation codebook that models all non-predictable components of the speech excitation signal. In the ACELP tool, a USAC block (e.g., 1,024 samples) is divided into four frames of 256 samples, which in turn are divided into four sub-frames of 64 samples (three sub-frames in the case of 768-length blocks).

The parameters of the long-term prediction filter and the innovation codebook are updated and transmitted at the sub-frame rate, resulting in a very high time resolution compared to the FD and TCX tools. Furthermore, the sparse algebraic code (SAC) structure in the ACELP innovation codebook has a resolution of 1 time-domain sample, allowing it to track events in the signal (e.g., attacks or onsets) very precisely in the time domain.

The LPC filters are quantized using the line spectral frequency (LSF) representation. To code the LSF, vector quantization (VQ) is used in a first-stage approximation followed by an optional algebraic vector quantized (AVQ) refinement. In the decoder, the LP parameters are converted to line spectral pair (LSP) representation and interpolated across sub-frames. The excitation parameters are determined in a closed-loop analysis-by-synthesis search procedure. The excitation pulse positions and gains are not transmitted, but rather a parameterized algebraic representation of the pulse sequence is transmitted.

Since TCX already incorporates a short-term prediction mode, incorporating the ACELP requires only a few additional tools, as shown in Fig. 8.5.

When incorporating the ACELP coding tools, it is important to maintain a smooth transition between ACELP and TCX or FD coding modes. ACELP frames (time-domain coding) apply a rectangular, non-overlapping window to the input samples, while FD and TCX frames (frequency-domain coding) apply a tapered, overlapping window to the input. As discussed in the previous section, FD and TCX both share the MDCT and they have a windowing scheme that permits seamless transitions

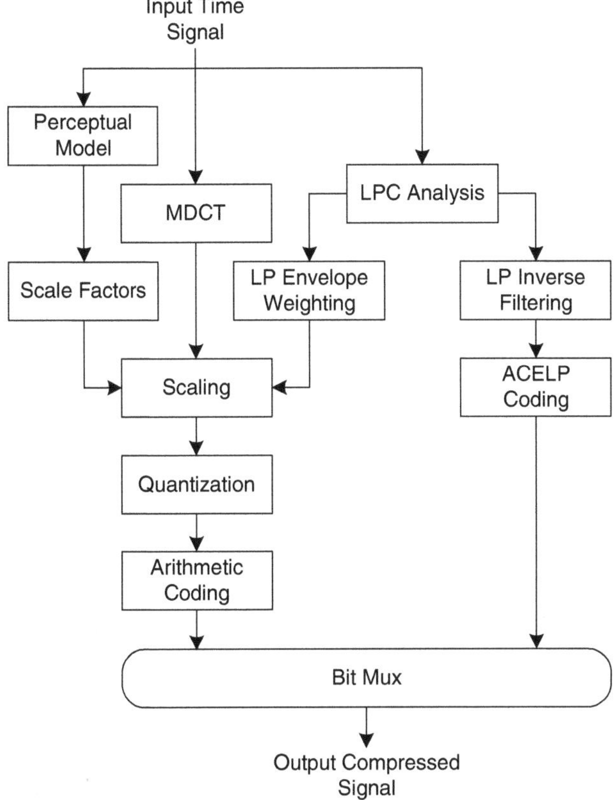

Fig. 8.5 Combined AAC, TCX and ACELP coding architectures

with full time-domain aliasing cancellation (see Fig. 8.4). The final step in ACELP coding mode is a short-term linear prediction synthesis filter that has no inherent means to cancel time-domain aliasing; furthermore, it has the state of the short-term predictor to contend with as the coder transitions from ACELP coding mode.

The solution to this problem is relatively simple – transmit at every transition between FD/TCX and ACELP a coded signal that represents exactly the time-domain aliasing cancellation [14] required by the adjacent MDCT. This would include FD long and short MDCT windows as well as the three TCX MDCT windows. This technique is aptly called the forward-aliasing cancellation (FAC).

8.2.2.4 Forward Aliasing Cancellation

Figure 8.6 shows the sequence of an ACELP frame followed by a TCX 512 frame followed by an ACELP frame. The four ACELP sub-frames are indicated by the small tick marks. Only the portion of the TCX window that is within the TCX frame

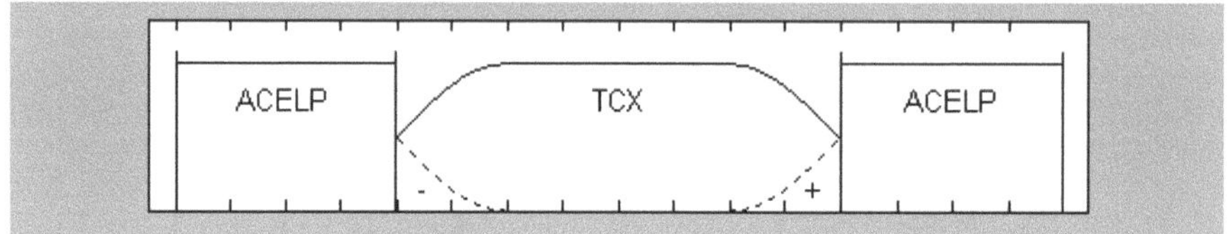

Fig. 8.6 ACELP to TCX transitions showing time-domain aliasing terms as dashed lines

is shown, as there is no neighbouring FD/TCX frames with which it can perform an overlap-add. Because of the lack of overlap-add, the un-cancelled aliasing is shown at each frame edge as dashed lines.

The encoder must determine the FAC signals that, when added to the left and right edges of the TCX frame, fully cancel the time-domain aliasing and bring the TCX envelope to unity magnitude.

In the decoder, assume that the leftmost ACELP frame and the TCX frame have been decoded. At the left side of the TCX frame, the last two sub-frames of the ACELP signal are windowed, time-reversed and used as a component of the FAC signal. In addition, the "zero-impulse response" (ZIR) of the ACELP synthesis filter that is valid at the last sub-frame is used a component of the FAC signal. The signal that must be added to these two components is the FAC signal that must be coded and transmitted.

The situation at the right side of the TCX frame is much simpler – the adjacent ACELP frame is not yet decoded and so is not known and there is no ZIR component. Therefore the entire FAC signal must be coded and transmitted.

If the center frame is coded with TCX (as shown), then noise shaping must be imposed on the FAC signal – at the encoder the FAC signal is filtered by the weighted linear prediction model W(z) prior to coding; at the decoder the FAC signal is filtered by 1/W(z) after decoding. The FAC procedure also applies to the case in which the center frame is FD (e.g. eight short windows), but with FD coded frames there is no need for noise shaping using the W(z) filter.

The FAC signals are coded by using a DCT-IV transform and quantized with the same algebraic vector quantizer (AVQ) used in LP filter coding.

8.2.2.5 USAC Architecture

We have reviewed the basic coding tools that determine the USAC architecture and provide the majority of the increase in performance when coding a mix of speech and music content at low bit-rates. A more detailed block diagram of USAC is show in Fig. 8.7. While previous figures were for an encoder, this figure is the USAC decoder. In fact, even this figure is simplified, but still presents the major coding tools that comprise USAC.

The following section will review some of the USAC coding tools that have not yet been discussed.

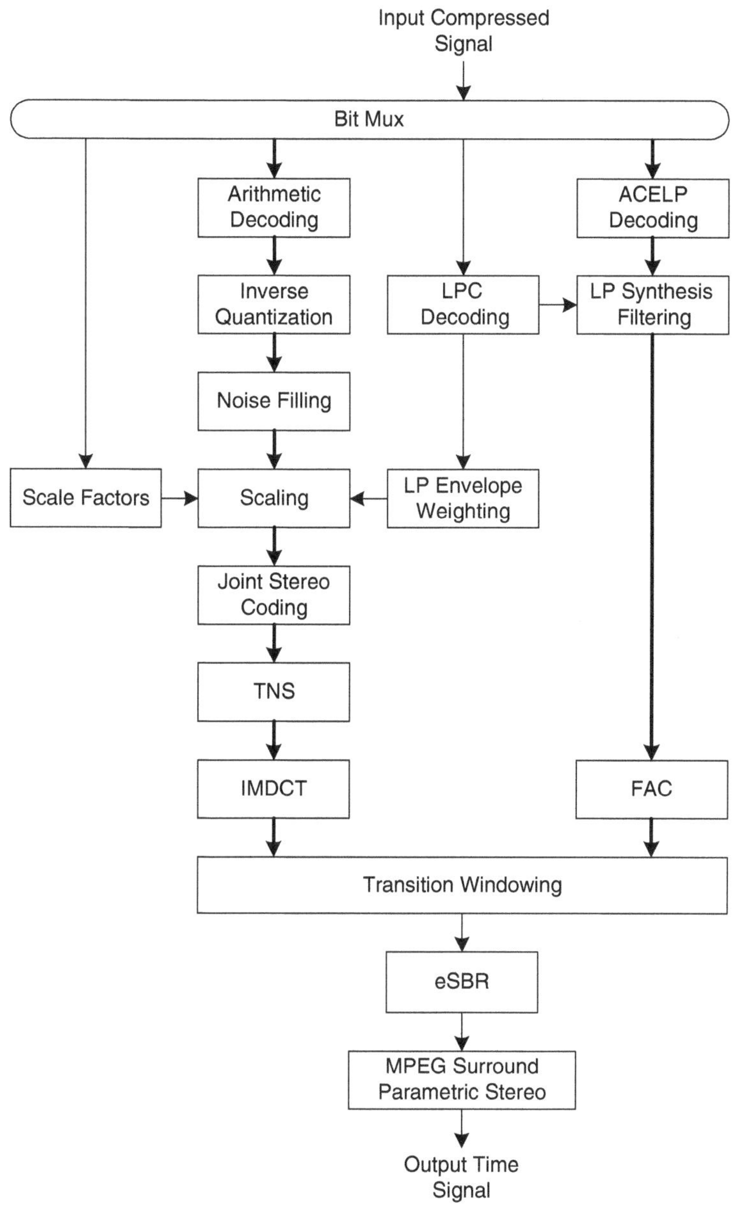

Fig. 8.7 USAC decoder block diagram

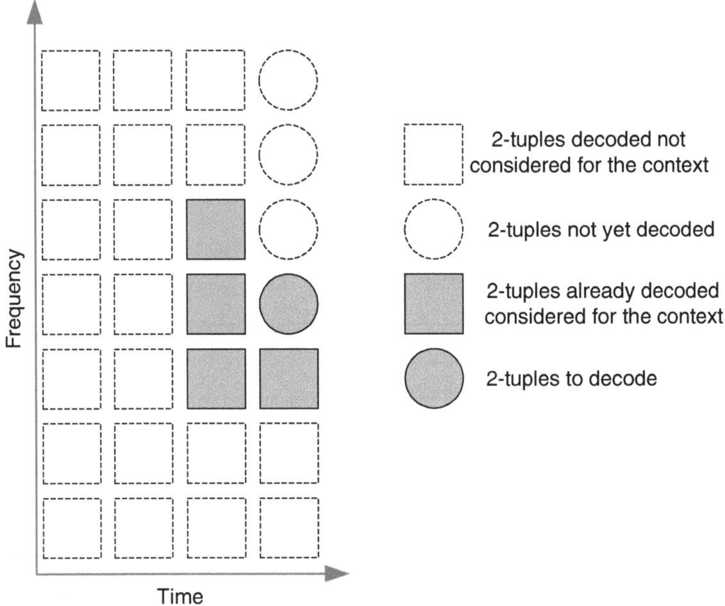

Fig. 8.8 Context-adaptive arithmetic coding

8.2.3 Description of Selected Coding Tools

8.2.3.1 Arithmetic Coding of Spectral Coefficients

MPEG AAC uses Huffman coding as the entropy coder for compressing quantized spectral coefficients. Since, as previously stated, there is no need to be backward-compatible with AAC, USAC is able to achieve a greater level of performance by replacing the Huffman coding with adaptive, context-dependent arithmetic coding. The spectral noiseless coding scheme is based on 2-tuples composed of two adjacent spectral coefficients. The coefficient pair is split into three tokens: the sign bits, the two most significant bits in each coefficient, and the remaining least significant bits. The signs are coded using one bit each whenever the quantized coefficient value is non-zero. The coding of the most significant two bits uses context-dependent cumulative frequency tables derived from four previously decoded 2-tuples, where tokens can be adjacent in either time or frequency, as illustrated in Fig. 8.8. The coding of the remaining least significant bits uses context-dependent cumulative frequency tables derived from the significance of the upper bits in the 2-tuple.

8.2.3.2 Noise Filling

The USAC noise filling tool is similar in concept to AAC's perceptual noise substitution (PNS) tool, but PNS fills noise at a given level for all spectral coefficients in a given scale factor band, whereas USAC noise filling permits an arbitrary mix of coded spectral values and artificially generated noise within a scale factor band. In the decoder, zero-valued coefficients above an offset frequency are replaced by random noise with a mean value equal to the mean quantization error in that scale factor band. Only one parameter representing the mean quantization error in the coding block needs to be transmitted, and in the decoder each band's noise energy is obtained by scaling this parameter by the band's scale factor.

8.2.3.3 Joint Stereo Coding

USAC has two tools for joint stereo coding. The first is Sum/Difference or Mid/Side (M/S) coding of channel pairs (i.e., left and right channel signals). This is done in the same manner as in MPEG-4 AAC [1].

The second tool is complex stereo prediction, which is a tool for efficient coding of channel pairs with level and/or phase differences between the channels. Using a complex-valued parameter α, the left and right channels are reconstructed via the following matrix, in which dmx_{Im} is the modified discrete sine transform (MDST) of the downmix signal and dmx_{Re} is the MDCT of the downmix signal.

$$\begin{bmatrix} l \\ r \end{bmatrix} = \begin{bmatrix} 1-\alpha_{Re} & -\alpha_{\mathrm{Im}} & 1 \\ 1+\alpha_{\mathrm{Re}} & \alpha_{\mathrm{Im}} & -1 \end{bmatrix} \begin{bmatrix} dmx_{\mathrm{Re}} \\ dmx_{\mathrm{Im}} \\ res \end{bmatrix}$$

The res and dmx_{Re} signals are transmitted to the decoder, where the dmx_{Im} signal is derived without the need for any additional side information.

8.2.3.4 Enhanced Spectral Band Replication (eSBR)

Relative to SBR in HE-AAC, USAC eSBR incorporates four major enhancements. First, there is a much greater range for the SBR crossover frequency which allows for greater flexibility in dynamically setting the crossover as the coder transitions through various coding modes. Second, the SBR up-sampling can be 1:4 (in addition to 1:2 as found in HE-AAC). The 1:4 up-sampling is appropriate when the core coder bit-rate supports only a narrow bandwidth. Third, the time/frequency grid for SBR parameters is more flexible and can have a finer time resolution which is particularly appropriate for the highly non-stationary speech signal. Fourth, eSBR has a new harmonic frequency transposition patching scheme.

Whereas spectral band replication (SBR) in MPEG-4 uses a patching scheme in which low-band time/frequency tiles are copied up to designated locations in the

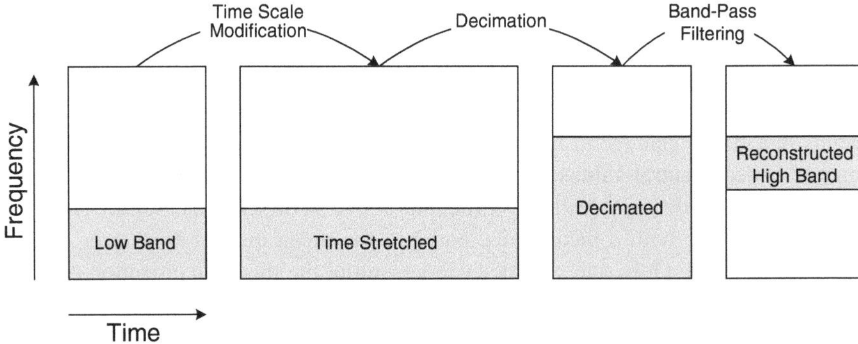

Fig. 8.9 eSBR harmonic transposition

high-band time/frequency space, harmonic transposition uses a short-time Fourier transform (STFT) to generate a "stretched" frequency version of the low band in which the continuity of the harmonic structure is preserved across the SBR cross-over frequency. In this respect, the STFT operates as a phase vocoder that performs time scale modification. The resulting signal is decimated and band-pass filtered to obtain a spectral segment that is used to reconstruct the high band. Transpositions are computed for factors of 2, 3 and 4 and a schematic version of the operations for a transposition factor of 2 is shown in Fig. 8.9.

After decimation, harmonics are on the correct "grid" but their frequency spacing is increased by the decimation factor. To obtain a dense set of harmonics, a method of filling in the harmonic structure from cross-products of the STFT coefficients is used.

The harmonic patching method typically gives better performance when coding music signals at very low bit-rates, in which the core codec may have limited audio bandwidth. This is especially true if these signals include a pronounced harmonic structure. On the other hand, the use of the regular SBR patching method is preferred for speech and mixed signals since it provides a better preservation of the temporal structure of the speech signal.

8.2.3.5 MPS212

USAC incorporates a subset of the functionality found in the MPEG Surround standard [2]; specifically it supports the 2-1-2 processing mode (i.e. mono to stereo synthesis) using the One-to-Two (OTT) box, shown in Fig. 8.10. The technology has been modified to make it more efficient when restricted to this one operating mode in the context of USAC. Therefore, it is referred to here as MPS212.

In MPEG Surround the spatial image of a multi-channel signal is coded into a compact set of time- and frequency-variant parameters – channel level difference (CLD), inter-channel coherence (ICC), and inter-channel phase difference (IPD). These provide a parametric representation of the human auditory cues for perceiving sounds in space [17].

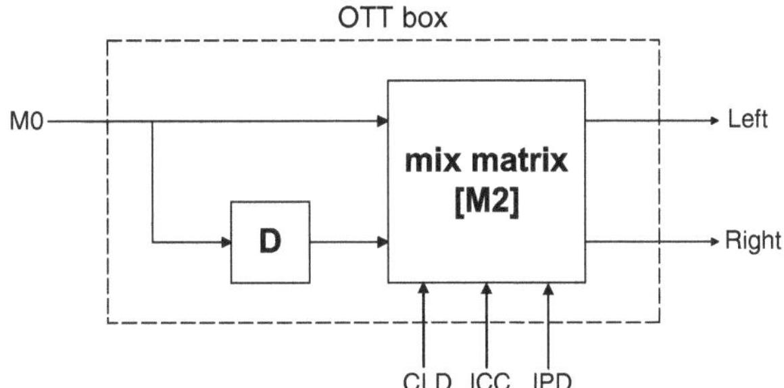

Fig. 8.10 One-to-two (OTT) box in MPS212 for mono to stereo synthesis

MPS212 is tailored to the needs of USAC in the following ways: first, the bitstream syntax is revised to maintain low overhead for the MPS212 parametric side information even when USAC is operated at the lowest bit-rates; second, the coding of the IPD parameter is improved; third, the complexity of the decorrelator (box "D" in Fig. 8.10) used in the mono to stereo synthesis is reduced while retaining the same level of quality; fourth, a new temporal steering decorrelator (TSD) is introduced; and fifth, a new mechanism of unified stereo coding is introduced.

The TSD tool provides a significant increase in performance for signals with very high spatial-temporal non-stationarity, such as applause or rain.

At higher bit-rates, when the eSBR and MPS212 modules may not be present, the core coder's "joint stereo coding" tool is used for stereo coding. However, this tool requires a full-band residual signal that permits complete time-domain aliasing cancellation in the IMDCT. At lower bit-rates, when the eSBR and MPS212 modules are present and the residual signal may not be full band, the MPS212 unified stereo tool is used to code the stereo signal.

The unified stereo tool efficiently represents a two-channel stereo signal by means of a downmix signal, a residual signal and MPS212 parameters. For this purpose, the stereo signal is decomposed into a sum and difference signal as follows:

$$l = s + d$$

$$r = s - d$$

The difference signal d is constructed from two components – one that can be predicted from the sum signal s and one, d_{res}, that cannot, as shown here:

$$d = -\alpha \cdot s + d_{res}$$

Since the coefficient α captures the signal power of the difference signal d that is coherent with the sum signal s, the resulting difference residual signal $d_{res} = d + \alpha \cdot s$ is orthogonal to the sum signal s. In general, α is a complex-valued prediction parameter that is determined from the MPS212 parameters CLD, ICC and IPD. These parameters are typically time- and frequency-varying, and the hybrid QMF domain representation of the signals is used to enable unified stereo processing to be done within QMF frequency bands.

Table 8.3 Test items used in CfP evaluation

Number of items	Content category
4	Speech
4	Speech mixed with music
4	Music

Table 8.4 Operating modes (bit-rate for mono or stereo coding) evaluated in CfP

	Bit-rate (kbps)	Stereo/mono
1	64	Stereo
2	32	Stereo
3	24	Stereo
4	20	Stereo
5	16	Stereo
6	24	Mono
7	20	Mono
8	16	Mono
9	12	Mono

Table 8.5 Systems under test

System	Description
USAC	USAC RM0
AMR-WB+	Reference codec
HE-AAC-V2	Reference codec

8.2.4 Performance

The report on the formal verification test on USAC is not yet available at the time of publication of this chapter. However, a good indication of the performance of USAC can be seen from the subjective test results associated with the Call for Proposals. The performance of every proposed system was evaluated for 12 test items when operating at nine modes (i.e., bit-rate for coding mono or stereo signals). The 12 test items represented a mix of speech and music content as show in Table 8.3. Speech mixed with music items consisted of both speech with background music as well as speech interspersed with short musical excerpts (e.g., “jingles”).

The nine operating modes evaluated are shown in Table 8.4. In this section only the performance of the best technology is relevant, as this technology, designated as reference model 0 (RM0) was selected as the basis for further development in the standardization process. In this discussion, the label USAC will be used to designate the performance of RM0.

In addition to being coded by all proposed systems, the items were coded by the two state-of-the-art speech and music coders – HE-AAC V2 and AMR-WB+ – as shown in Table 8.5.

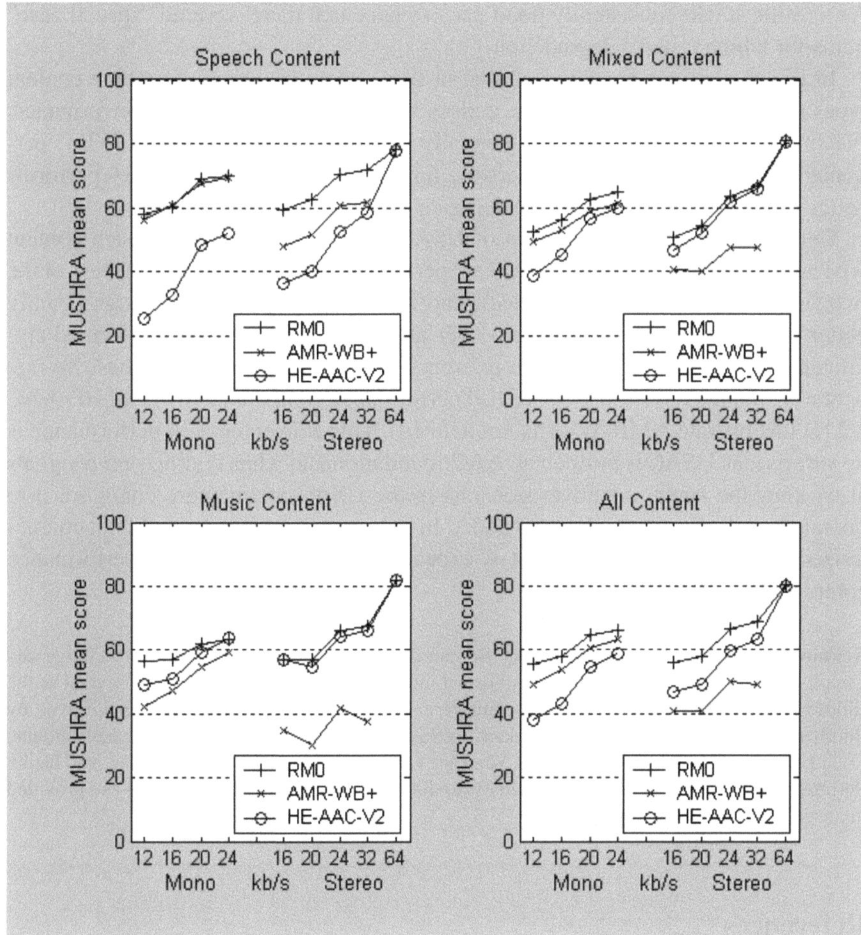

Fig. 8.11 Results of the CfP subjective test showing performance for USAC, reference codecs and anchor conditions

The subjective evaluation was done using the MUSHRA methodology [18]. The listening test was carried out at several test sites worldwide, and when pooled over all sites there were at least 30 listeners for each operating point. At 64 kbps there was no AMR-WB+ system since it does not support this operating point. The results of the test are shown in Fig. 8.11. All scores were reported as mean values, where the mean is over all subjects and all test items in the given content category.

One very important outcome of the test which is not shown in the figure, is that USAC had a mean subjective score at each operating point for each test item (as estimated by the 95% confidence interval) that was comparable to or better than the best of HE-AAC or AMR-WB+ for that operating point and test item. In other words, USAC was always as good as or better than the two state-of-the-art codecs.

As a result, it had consistently good performance and there were no "special case" items for which it had a degradation.

In Fig. 8.11 it can be observed that at some operating points for some content types the state-of-the-art reference codecs have a significant drop in performance. For music signals, HE-AAC V2 generally performs better than AMR-WB+, particularly for stereo signals. Conversely, for speech signals, AMR-WB+ performs better than HE-AAC V2 for the majority of operating points.

Comparing these results to that of USAC, it can be seen that for each content type and at each operating mode, USAC performed at least as well as the best of the two state-of-the-art reference codecs, and in many cases USAC had significantly better performance. When averaged over all content types, USAC performed significantly better than either of the reference coders for all but the highest bit-rate operating point. At 64 kbps, USAC had performance comparable to that of HE-AAC V2 (at this bit-rate AMR-WB+ is not defined). This convergence in performance is no surprise, as USAC is built on an AAC foundation and when coding stereo signals at 64 kbps the signal-adaptive decisions in the USAC architecture configure it to operate essentially as an HE-AAC coder. In this respect, although it was not characterized, at even higher bit-rates it is expected that USAC will have performance comparable to that of AAC.

Acknowledgements The author thanks Bernhard Grill of Fraunhofer Institute for Integrated Circuits for providing to the author unpublished information on USAC from which several of the slides in this chapter were adapted. The author also gratefully acknowledges the help given by Bernhard Grill, Max Neuendorf and Markus Multrus of Fraunhofer Institute for Integrated Circuits; Roch Lefebvre and Philippe Gournay of Voiceage Corp. and University of Sherbrooke; and Heiko Purnhagen and Kristofer Kjörling of Dolby who did a careful review of the draft text and provided many constructive comments.

References

1. ISO/IEC 14496–3:2009, Information technology – Coding of audio-visual objects – Part 3: Audio, Edition: 4.
2. ISO/IEC 23003–1:2007, Information technology – MPEG audio technologies – Part 1: MPEG Surround; Edition: 1.
3. ISO/IEC 23003–2:2010, Information technology – MPEG audio technologies – Part 2: Spatial Audio Object Coding (SAOC), Edition: 1.
4. ISO/IEC 23003–3:2011, Information technology – MPEG audio technologies – Part 3: Unified speech and audio coding, Edition 1.
5. M. Neuendorf et al., "Unified Speech and Audio Coding Scheme for High Quality at Low Bit-rates, ICASSP '09, Taipei, Taiwan.
6. M. Neuendorf et al., "A Novel Scheme for Low Bit-rate Unified Speech and Audio Coding – MPEG RM0," 126th AES Convention, Munich, Germany, May 2009. Preprint #7713.
7. ISO/IEC WG11 N9519, "Call for Proposals on Unified Speech and Audio Coding".
8. ETSI TS 126 290 V9.0.0 (2010–01), Digital cellular telecommunications system; Universal Mobile Telecommunications System (UMTS); LTE; Audio codec processing functions; Extended Adaptive Multi-Rate - Wideband (AMR-WB+) codec; (3GPP TS 26.290 version 9.0.0 Release 9).

9. Marina Bosi, Karlheinz Brandenburg, Schuyler Quackenbush, Louis Fielder, Kenzo Akagiri, Hendrik Fuchs and Martin Dietz, "ISO/IEC MPEG-2 Advanced Audio Coding," JAES Volume 45 Issue 10 pp. 789–814; October 1997.
10. Martin Dietz, Lars Liljeryd, Kristofer Kjorling and Oliver Kuntz, "Spectral Band Replication, a Novel approach in Audio Coding," in 112th AES Convention, Munich, Germany, May 2001, preprint 5553.
11. Martin Wolters, Kristofer Kjorling, Daniel Homm and Heiko Purnhagen, "A closer look into MPEG-4 High Efficiency AAC," 115th AES Convention, New York, Oct. 2003, preprint 5871.
12. Lefebvre, R.; Salami, R.; Laflamme, C.; Adoul, J.-P.; "High quality coding of wideband audio signals using transform coded excitation (TCX)," in Proc. ICASSP-94, vol. 1 pp. 193–6, April 1994, Adelaide, Australia.
13. Jari Makinen, Bruno Bessette, Stefan Bruhn, Pasi Ojala, Redwan Salami and Anisse Taleb, "AMR-WB+: a new audio coding standard for 3rd generation mobile audio services," in Proc. IEEE ICASSP '05, March 2005, vol. 2, pp. 1109–12.
14. John Princen and Alan Bradley, "Analysis/Synthesis Filter Bank Design Based on Time Domain Aliasing Cancellation," IEEE Trans. ASSP, vol. 34, no. 5, pp. 1153–61, 1986.
15. Laflamme, C.; Adoul, J.-P.; Su, H.Y.; Morissette, S., "On reducing computational complexity of codebook search in CELP coder through the use of algebraic codes," Acoustics, Speech, and Signal Processing, 1990. 1990 International Conference on ICASSP-90, vol., no., pp.177–180 vol.1, 3–6 Apr 1990.
16. Cox, R.V.; Kroon, P.; "Low bit-rate speech coders for multimedia communication," Communications Magazine, IEEE, vol.34, no.12, pp.34–41, Dec 1996.
17. Herre, Jürgen et al., "MPEG Surround-The ISO/MPEG Standard for Efficient and Compatible Multichannel Audio Coding," JAES Volume 56 Issue 11 pp. 932–955; November 2008.
18. ITU "Method for the subjective assessment of intermediate sound quality (MUSHRA)," 2001, ITU-R Recommendation BS. 1543–1, Geneva, Switzerland.

Chapter 9
MPEG System Basics

Peter P. Schirling

9.1 Organization of this Chapter

This chapter is intended as a mid-level view of MPEG systems for those who wish to gain an understanding of the concepts. This chapter will provide enough detail to permit the reader work with those who are implementing an MPEG system. The reader should consider having a copy of the current specification handy while reading this chapter but this is not necessary in order to grasp the concepts that will be presented.

9.2 Purpose

Why is MPEG-2 Systems needed to begin with? The systems layer, as it is referred to, provides the means to identify information carried in the stream such as audio, video and the other data needed in order to make the collections of bytes in the stream intelligible. In a broadcast application, such as terrestrial systems or cable TV applications it also provides the means to deliver the clock necessary for synchronizing the audio and video. Transport Streams can carry multiple programs. Part of the Transport Stream multiplex contains the information to identify which elements of the stream contain audio and video that belongs together. The other stream format, Program Stream, is used primarily in applications where programs are stored such as DVDs. These are far less prone to delivery errors than broadcast. Program Stream syntax varies from Transport Stream but it performs the same functions. Program Streams can only carry one program.

P.P. Schirling (✉)
Lakewood Consulting, LLC Burlington, Vermont US
e-mail: vttrainman@ieee.org

L. Chiariglione (ed.), *The MPEG Representation of Digital Media*,
DOI 10.1007/978-1-4419-6184-6_9,

9.3 Principles

Unlike symmetrical and interactive multimedia systems available today, MPEG-2 was developed for broadcast systems that are asymmetrical and passive. This paradigm implies (a) that there are millions of decoders (TVs) for every encoder (transmitter) and (b) the receiver cannot control the actions of a transmitter.

It was also a time when the cost of components (memory and logic) was a far greater factor than it is today. This made it necessary to implement the early (real-time) MPEG-2 encoders and decoders in circuit logic specific to the MPEG-2 requirements. Likewise, the size of the memory required to implement the buffers called out in the specification of the MPEG-2 Systems layer were considered very carefully and limited to only that needed to fulfill the exact requirements called for in their respective specifications. It will be seen that these factors drove many of the details associated with the MPEG-2 specification overall and MPEG-2 Systems specification specifically. Even though current MPEG-2 platforms are able to use general purpose hardware and software with large amounts of memory to implement MPEG-2, as well as other AV specifications available to today's consumer, the repetitive nature of transmitting the header necessary to interpret an MPEG-2 stream would still require careful consideration.

There are fundamental technical principles behind MPEG-2 streams as well. Streams must be delivered "on time" and error free. The systems reference model described in a later section will speak to this notion of on time in a little more detail. Implementing an MPEG-2 systems platform requires the detailed reading and understanding of the System Target Decoder (STD) models. It is possible to implement an MPEG-2 systems platform as an exact replication of an STD, though a clear understanding would lead a designer to a possibly more pragmatic and efficient implementation.

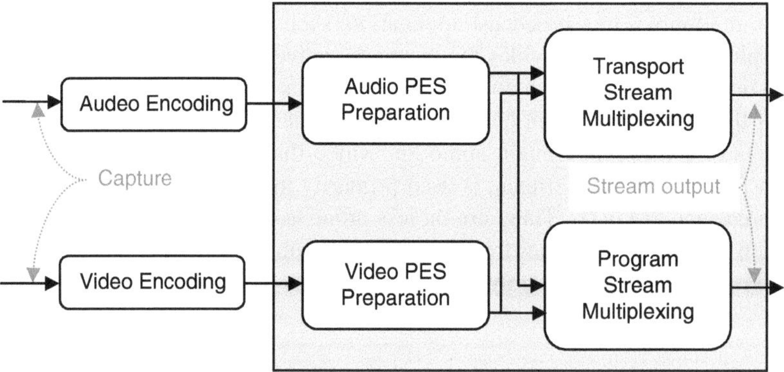

The MPEG-2 Systems specification does not specify an entire end-to-end system but, as the shaded area of the diagram above illustrates, it only specifies the bit stream format and how to assemble the component parts of the stream into either a Transport Stream or Program Stream.

The notion of data arriving error free is clear. In reality both broadcast and stored data environments can deliver information that still contains errors if the front end electronics (e.g. tuners) cannot correct them. The impact of these errors varies. Corruption of information in a header data is likely to have a greater impact on the operation of a decoder platform than bad bits in the compressed layers of audio or video. This chapter will not discuss the impact or the methods used by network providers to detect and correct errors caused during the delivery of bit streams.

In general, broadcast networks and especially terrestrial broadcasting do not have end-to-end network clocks. Therefore, an accurate clock needs to be part of the bit stream in order for the receiver, whose clock is a slave to the broadcasters clock, to be able to synchronize the audio, video and any associated data that must also be synced to either the audio or the video, such as closed captioning. If the bit stream is delivered accurately the receiver will use a phase locked loop to control its frequency to match that of the transmitter. If the receiver's clock is allowed to drift to frequencies higher or lower than the transmitter's the rate at which audio and video decode will be too high or too low and will ultimately cause noticeable audio and video events. The reason is that the buffers that hold compressed audio or video will be depleted at an incorrect rate causing either buffer underflow or overflow.

An extension of this principle of "on time" delivery is that MPEG streams have a constant end-to-end delay. Simply stated, any 2 bytes that left the multiplexor separated in time, for example, by 1 ms must arrive at the receiver separated by 1 ms. The delay can be computed by adding up the times associated with encoding, buffering the encoded data, and then multiplexing it into one of the specified formats. The delay will also include any transmission (e.g. terrestrial broadcast) or storage (e.g. DVD or video server) and then the inverse processes of demultiplexing, buffering, decoding and then displaying the content. Also, included in this notion is the concept that unless otherwise indicated by certain fields in the bit stream, each picture or audio sample is presented only once. This is an important concept because the bit rates of audio and video vary greatly yet each must arrive at their respective decoders such that they can be decoded and rendered one at a time continuously as well as synchronized correctly. This concept is defined as a rate division multiplex.

9.3.1 Broadcast

As stated earlier, the Transport Stream is used primarily in broadcast applications that are prone to errors from a variety of mechanisms that will not be described here. Therefore, the basic element of a Transport Stream, the TS Packet, is a short 188 bytes to allow for the use of ECC (error correction codes). As an example, the satellite delivery systems have defined a super frame of 204 bytes for the addition of 16 bytes of ECC. The diagram below illustrates such a system.

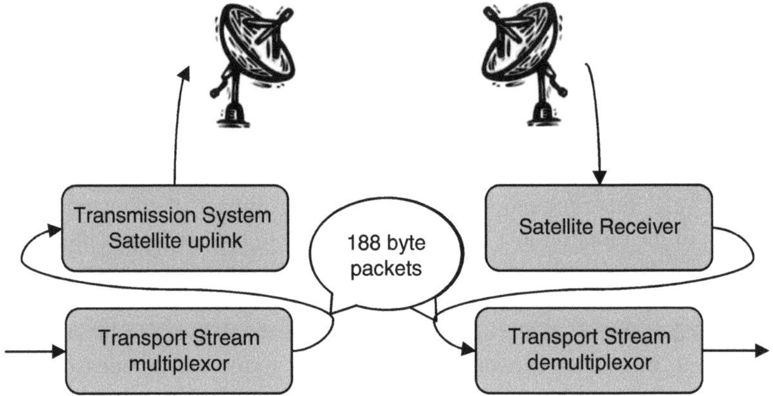

Note that the addition of those bytes in the stream must be accounted for in the bit rate of the transmitted multiplex in order to ensure that the MPEG-2 stream arrives at the demultiplexor at the correct bit rate after the 16 bytes of forward error correction (FEC) are removed. This is true of any wrapper that might be used to deliver TS or PS packets.

One of the key features of MPEG-2 Transport Stream is that it can be re-multiplexed. This means that programs from one Transport Stream can be extracted or deleted. Likewise, a Transport Stream can be added to another Transport Stream. This will be described in a later section. The illustration below is an example of Transport Stream remultiplexing. It shows both the addition and deletion of a stream to a multi-program Transport Stream.

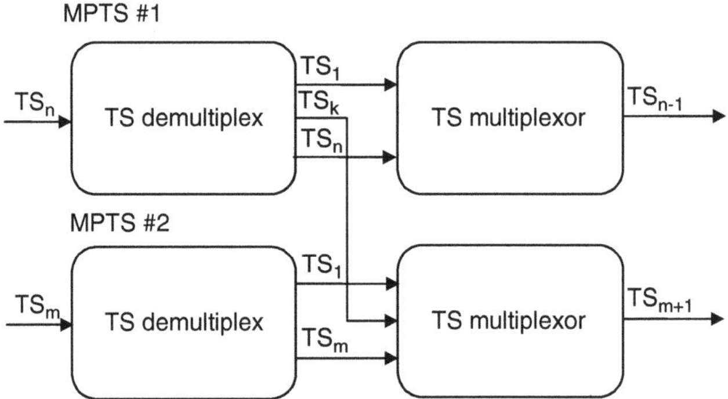

9.3.2 *Historical Note*

Why 188 bytes? Two reasons, the first being that during the development phase of MPEG-2 systems it was discovered that multiple fixed TS packets lengths could not by re-multiplexed and thus a single fixed length packet was required. The length of

188 bytes was selected to accommodate ATM adaptation layer 2(AAL2). The payload of an ATM datagram is 47 bytes ($4 \times 47 = 188$). At that time ATM AAL2 was thought to be the means of fiber-to-home delivery for digital video. History has proven otherwise.

9.3.3 *Stored Data Applications*

Stored data applications, on the other hand tend be far less error prone and thus Program Stream allows for longer data unit lengths. Program Stream is also backwards compatibility with MPEG-1 SysMux. This was a requirement for the MPEG-2 systems specification. DVDs, including the latest generation Blu-ray™ discs use Program Stream.

9.4 Stream Types

As already mentioned there are two stream types defined by the MPEG-2 systems standard to carry audio, video and all the associated data needed to consume the bit stream. The following will explain the basics of Transport Stream and Program Stream.

9.4.1 *Transport Stream*

The 188 bytes of each Transport Stream (TS) packet are divided into two layers, the systems layer and the compression layer or payload as it will be referred to further down in the text. The systems layer is itself divided into two parts as well. The first 4 bytes of each TS packet are always present. The remaining bytes of a TS packet can either be additional systems layer data that is referred to as application fields. The remaining Transport Stream bytes after that can be payload or stuffing bytes. This is signaled in a field carried in the first 4 bytes.

Before the bits and bytes of a TS packet are described it is important to have an understanding of the System Target Decoder (STD) model. The Transport Stream STD (T-STD) and the Program Stream STD (P-STD) described later, are the means by which an MPEG-2 systems platform are implemented. The T-STD is a mathematical model that describes the arrival time, buffering and redistribution of each byte and collection of bytes to its associated decoder. While the bytes carried in the stream are essential to the understanding and ultimately the presentation of audio and video, the T-STD describes how each byte is treated as it arrives at the input of the demultiplexor. The pacing of the decoding platform is controlled by the systems layer and specifically by the clock (Program Clock Reference) carried in the stream. Hence, the requirement for the on time arrival of each byte at the demultiplexor input.

The value carried in the PCR controls the System Clock Reference (SCR) frequency that in turn controls the pace at which video and audio are decompressed and presented. The SCR is common to both the video and audio decoder. Correct pacing will ensure that the audio and video are synchronized when displayed.

The notion of on-time arrival and constant delay is embodied in the STD model. Buffering and redistribution of the bytes received is also described by the STD. As stated previously, it is an idealized mathematical construct that can be implemented exactly as described or through careful study it is usually implemented in a more pragmatic and efficient manner. It is recommended that the T-STD and P-STD are studied and understood in detail when attempting to design and implement a systems layer platform.

A Transport Stream is made of continuous bytes assembled into packets that must be identified in order for the bytes to make any sense. The diagram below illustrates how a Transport Stream is assembled.

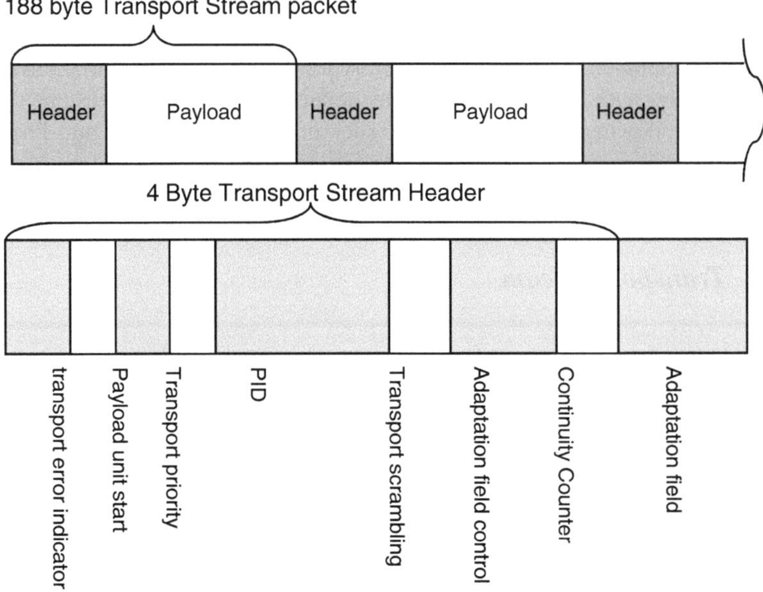

The first byte of each TS packet is the sync byte. It has a hex value of 0x47. During the development phase of MPEG-2 it was determined that this combination of bits would permit a receiver to sync to an incoming bit stream in the shortest possible time.

When a receiver is turned on, whether it is a terrestrial receiver, a cable set top box or a satellite receiver it must first tune to a carrier frequency and then begin detecting bits. This process is not part of the MPEG-2 specification. Once the tuner can begin to discriminate individual bits in the stream the demultiplexor can begin to search for the sync byte. Once three successive sync bytes are detected separated by 188 bytes the bit stream can be considered to be byte synchronized. Acquisition and decoding of TS packet information can now begin.

The next 2 bytes, starting at the high order bits, of all TS packets contain the following elements. A single bit that can be set by the front end tuner electronics to indicate if this packet contains one or more data errors that could not be corrected before being handed off to the systems layer demultiplexor. Note, that if the sync byte has an error, TS sync cannot be achieved or maintained and the acquisition phase will continue or restart. MPEG does not require that this bit be set or specify what to do if the bit is set to "1".

The next bit indicates if the payload portion of this packet is the first byte of either a PES (Packetized Elementary Stream) or PSI (Program Specific Information). PES packets carry the compressed audio video data while PSI data contains the information about the programming carried in a multiplex. These stream types will be explained below.

It was explained earlier that errors in the bit stream will have an impact on the receivers depending on the location of the error. The next bit, which can be set at the multiplexor, can be used by transmitters, to indicate that this packet contains data that is more important than other packets and care or action could be taken to ensure its proper delivery. MPEG does not specify whether this bit is required to be set or how to set it or even what action can be taken to use it.

The next 13 bits, the last of these 2 bytes, are critical to the successful operation of a receiver platform. The PID (Program Identification) field is used by the demultiplexor to recognize and then separate audio streams from video streams from the other data and information carried in the multiplex. Recall that one of the features of Transport Streams is the ability to deliver multiple programs so the PID is crucial in being able to associate streams of audio and video belonging to the same program. Incidentally, a program is defined as all audio, video and data with a common PCR (Program Clock Reference). The PID is also used to identify the content of the multiplex as specified in the PSI (Program Specific Information) tables.

The next byte is the last byte of the always present 4 bytes of a TS packet and it contains 3 fields. The high order 2 bits identify whether any of the remaining bytes of the payload are scrambled (encrypted). The first 4 bytes cannot be delivered to the demultiplexing platform encrypted for obvious reasons. The four combinations of this 2 bit field simply indicate whether there is no scrambling or there is scrambling (encryption). If scrambling is used, MPEG does not specify which of the remaining bytes that are allowed to be scrambled are encrypted or what methods are needed to decrypt them. This is left to the implementer and agreements between broadcasters and the manufacturers of receiving equipment. Other information related to scrambling (encrypting) is provided in section on PSI (Program Specific Information).

The next field is also 2 bits and it specifies whether there is more TS packet data or if the next byte and those following contain other information as indicated by the combination of bits in the field. One value in the field is reserved. The remaining three possibilities indicate if there is only payload data, only adaptation field data and no payload or lastly both adaptation field data and payload.

The last 4 bits of the 4 byte TS packet header is a continuity counter. It can be used to determine if one or more TS packets have been lost during transmission. It can also be used when data considered critical to an application and the same TS packet is sent twice in consecutive packets. This is not a deterministic method of detecting lost packets since it is possible to lose 15 packets and thus the next packet received has what appears to be a correct continuity counter value but is, in fact, not intended to be a redundant TS packet. Also, the continuity counter only increments when there is adaptation field information carried in the TS packet. Additionally, when the adaptation field of a redundant packet contains a clock value (PCR), the value carried in the PCR field is updated to reflect a correct arrival time.

The remaining 184 bytes in a Transport Stream packet are either adaptation field bytes or payload bytes or a combination of these. If there is insufficient information to fill out the 188 bytes the remaining bytes are filled with stuffing bytes whose value is always 0xFF (all ones).

The first byte in the adaptation field signals the total number of bytes that follow. If there is only adaptation field data the value can be in the range of zero to 183. If there is payload data then the adaptation field can have a maximum length of 182 bytes since at least 1 byte of the TS packet has to contain payload data.

The second byte of the adaptation field is made up of individual bits used to indicate if a particular field is present in subsequent bytes.

The first (high order) 3 bits of this second byte are used to signal that the payload of current or subsequent TS packets have special meaning or information not contained in other payload data. One of the bits deals with the real world of commercial broadcasting where programming is continuous and has commercials or material originates from different sources. These switches cause what is known as discontinuities and signaling them in the bit stream is essential to smooth operation of a decoder platform and thus avoids a poor viewer experience.

The next bit is used most often in stored applications when the viewer wants to fast forward (or reverse) through a program. If set by the broadcaster, this random access indicator can be used by a demultiplexor to skip to the start of a frame further ahead (or back) in the stream. If it is used the PES packet and picture start codes are usually synced in the TS packet that carries them to make searching and subsequent decoding quicker.

The third bit is a video elementary priority indicator and its use is restricted to the presence of specific video coding data. To better understand the use of these bits it is best that the reader consult the MPEG-2 systems specification for details.

The next bit signals the presence of the PCR (Program Clock Reference) value in the 6 bytes that follow. The PCR is the critical element that enables successful operation of the receiver platform. The PCR is a slave to the broadcaster's system clock and when properly controlled ensures synchronization of the audio and video of a program to its source. The PCR is embedded in the stream at the time the stream leaves the broadcast multiplexor. Recall that transport streams can carry multiple programs. In general, a broadcast multiplex begins life as a single program Transport Stream (SPTS) and the values carried in the PCR reflect the correct delivery time of that SPTS. However, the advent of digital TV provided broadcasters with the ability to embed multiple programs encoding onto the same carrier channel. When an SPTS is remultiplexed into a multi-program Transport Stream (MPTS) the delivery time of the PCR, as well as all the other bytes, embedded is shifted in time. Therefore, in order to maintain proper operation of the decoder platform when this program is played the values carried in the PCR field will be adjusted by the remultiplexing system to reflect the shift in delivery time caused by remultiplexing.

Certain types of scrambling applications however, require the data used during the initial scrambling to descramble the bytes that follow. MPEG added a field called the OPCR or original PCR as a means of permitting this condition. State-of-the-art encryption technology does not require original data so the use of the OPCR field is infrequent at best.

The PCR (and OPCR) fields are in two parts as illustrated in the diagram below.

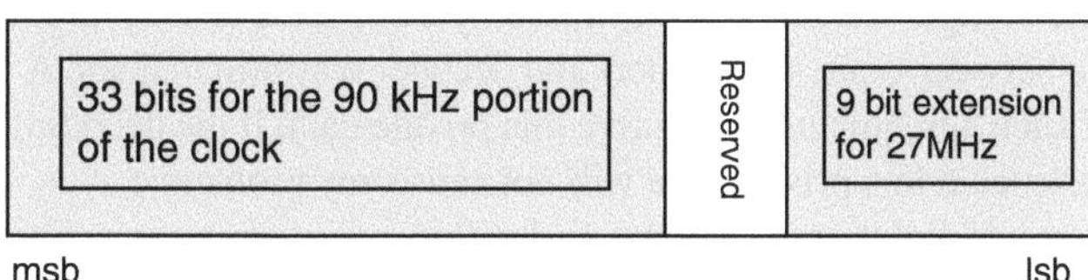

The high order 33 bits provide for a 90 kHz clock. This 90 kHz clock is backwards compatible to the MPEG-1 90 kHz clock used to maintain correct operation of MPEG-1 streams. However, the bit rates of MPEG-2 streams, especially MPTS, is far greater than those of MPEG-1 and thusly a 9 bit extension was added to provide a 27 MHz clock for MPEG-2. Note, that the 9 bit extension is carried in the lowest 9 bits of the 6 byte (48 bit) field. The PCR and its associated SCR allow for times that exceed 24 h of wall clock time. Operation of the clock is such that each time the 27 MHz portion (low order 9 bits) overflows the low order bit of the 90 kHz clock is incremented by one.

Synchronization of time dependent streams in a Transport Stream multiplex is fundamental. It is the frequency of the System Clock Reference (SCR) that is critical to synchronization of the decoding platform. The MPEG-2 Systems specification requires that PCRs appear in the bit stream no less often than every 700 ms. The initial value of the decoder SCR is set when the first PCR arrives after the Transport Stream is synced. Subsequent values of the PCR are used to control the frequency of the SCR generally using phase locked loop circuitry. Allowing the frequency to drift will cause visible and/or audible effects because the individual decoders pace themselves off the SCR. If the frequency is too high the buffers holding compressed data will empty too fast and will quickly have no compressed data to work with. If the frequency is too low then decoding will also not match the broadcaster's timing and the buffers will deplete too slowly and quickly overflow, again causing undesirable events. In both cases recovery is intrusive because it requires the receiver to resynchronize and reinitialize decoding process. While not described in the T-STD of the MPEG-2 system's specification, implementers should also take care not to over control the SCR frequency as this will also cause platform control issues.

One of the bits in the byte following the adaptation field length indicates if there is going to be a discontinuity in the PCR value carried in this PID stream, such as when commercials are inserted into a broadcast stream. The value that is carried in this field is a signed value. A positive value indicates the number of packets until the change occurs. A negative value indicates the number of packets since the change. Other aspects of the rules associated with this field can be read in the specification.

MPEG standards specify the information necessary for successful consumption by their associated decoders of the bytes contained in their respective bit streams. There are occasions when broadcasters have information that is unique to a program for which there is no place in the bit stream for it to be carried. MPEG provides a place for such information called private data. A 1 byte field plus its associated data can be part of the adaptation field. Two hundred and fifty six bytes are theoretically possible. However, this data cannot span Transport Stream packets. Therefore, the actual maximum length of such data is 180 bytes. This private data can only be used by decoders with prior knowledge of the content of these bytes. This is usually accomplished by an agreement between content providers and decoder platform makers. A decoder that is not part of such agreements will ignore this private data. If the decoding platform is properly constructed this private data will not cause any problems.

The next set of fields is a complex collection of information related to stream management and is best understood by reading the specification.

As explained earlier, if there is insufficient information to fill the 188 bytes of a TS packet the remaining bytes are filled with stuffing bytes whose value is always all “1”s. Stuffing bytes are ignored by compliant platforms. However, some network operators insert “stuffing packets” with private data to perform network diagnostics. The network operator’s premise equipment is built to utilize this data when carried in the stream and utilizes an uplink to report the information it gathers. This is implementation specific and, like private data, the MPEG-2 specification is silent.

This concludes the Transport Stream packet header information. The audio and video data are carried in a stream of bytes referred to as PES packets (Packetized Elementary Stream) packets. Recall that carrying payload data in a TS packet is signaled using the Adaptation Field control bits and that the start of a PES packet is signaled by a bit in the byte immediately following the sync byte.

9.4.2 PES Packets

The syntax of Packetized Elementary Streams (PES) packets is common to Transport Stream and Program Stream.

PES packets begin with header information. The PES packet header is a collection of flags and fields not all of which will be described in detail here.

The first 24 bits (3 bytes) are a legacy of MPEG-1 PES packet headers. Called the start code prefix, it is always 23 zeros followed by a one. The diagram below highlights only the initial bytes of the PES header.

Start code prefix	Stream ID	Packet length	Flags for more PES HDR data	PES data bytes

The next byte (Stream ID) identifies the type of stream contained in this PES packet. Only one type of stream, audio or video or other stream type identified by a table in the specification can be carried in PES packets with a specific byte value. The next 2 bytes specify the total length of this PES packet. However, video stream types carried only in Transport Steams are a special case in which the length field can be set to zero. This exception was made for live program distribution, to allow a video PES packet to leave the video encoder before its length could be inserted in the PES packet header. This reduces the delay between encoding and ultimately when the compressed stream and the multiplex can be transmitted.

The next 2 bytes contain flags that signal the presence of additional data in the PES packet header. The first of these indicates whether the compressed data in the PES packet is scrambled. Again, like TS packet headers, PES packet headers can never be scrambled.

The next 2 bits are used by the video encoder to inform the MEG-2 Systems multiplexor that this PES packet is more important than other PES packets. The multiplexor can use this to set its priority indicator but the multiplexor cannot change the bit back to a zero. The next bit tells the decoding platform that the first data byte of payload is the start code for video or audio. There is a complex relationship between this flag and other data in the stream.

The copyright bit and the following bit are related to what is called DRM (Digital Rights Management). The specification details their use in a bit stream.

The following 2 bits indicated whether there are one or both timestamps associated with decoding compressed data. MPEG-2 video data does not have to be sent in the order in which it is displayed. However, the data needs to be decompressed and held for use by subsequent compressed data before it is displayed. The PTS (Presentation Time Stamp) and DTS (Decoding Time Stamp) provide the necessary information to do that. These fields are also used to pace the decoders.

Next are the Elementary Stream Clock Reference (ESCR) and the Elementary Stream rate fields that are part of the synchronization mechanism at the PES layer and used primarily for stored applications.

Stored applications including the class of receivers that store broadcast streams can use something called trick mode. These are familiar controls like fast forward, freeze frame, etc. This collection of fields enables the use of trick modes. The random access bit is part of this capability.

A second set of private copyright information can be included in the PES header when the flag is set indicating its presence.

Another means of checking whether data has arrived error free, this next field contain the CRC (Cyclic Redundancy Check) code for the previous PES packet. The PES header is not included in this because PES header data can be altered during transmission.

The PES header can become very large so flags and fields are included to signal whether additional data is present or not. One of those additional fields carries a Pack Header. This data enables the reconstruction of a Program Stream from a Transport Stream. There are numerous other fields that are best understood by reading the specification.

9.4.3 Program Specific Information (PSI)

As stated earlier, Transport Streams can carry one or more programs. In order to distinguish one program from another, video from audio from data within a program, it is necessary to carry information in the stream that serves this purpose. PSI data does just that. In fact, acquisition of the PSI tables is the first thing to occur once Transport Stream sync is achieved. PSI tables are signaled via PIDs.

The Program Association Table is the root table that must have a PID value of zero (0×00). Only a few other PIDs have a fixed value. These include the Conditional Access tables used to manage the scrambling aspects of the Transport Stream, the Transport Stream Description Table that describe transport wide information, and the IPMP (Intellectual Property Management and Protection) tables. The assignment of the other PID values is the purview of the network operator.

There are two types of tables that will be used in specifying the content of a Transport Stream multiplex. Program Map Tables specify the PID values for each of the components of a program being carried in a multiplex. The syntax of these tables is specified by MPEG. Network Information Tables carry physical network information about the network, such as transponder numbers in the case of satellite delivery systems. The syntax of these tables is not specified by MPEG.

A feature of PSI tables is that they can specify future programs that will be available on the provider's network. PSI tables do not necessarily need to be carried in the stream to which they refer. Some providers send their entire collection of programs in a stream dedicated to PSI information. For example, MSOs (cable operators) send their program information to a set top box even though the consumer may have turned it off. The advent of DVRs has also made the availability of future programming an important aspect of consumer satisfaction.

Unlike audio/video streams, PSI table information is not compressed and most importantly it is sent repeatedly although, unlike PCRs, MPEG does not specify a periodicity. When a receiver is turned on if the PSI tables are not already present, as explained earlier, they can then acquire the information needed in order to distinguish the audio, video PIDS and other PIDs that may carry information associated with a program.

Once PID zero is detected in the Transport Stream it is still necessary to locate the first byte of the PAT (Program Association Table). Finding the first byte of a PSI table is accomplished by what is called the pointer field. Each TS packet carrying PSI tables has the pointer field to a value that is the offset in payload where that first byte is located. If no section begins in this TS packet then the payload start indicator is not set and no pointer value is present.

Note that all PSI tables start with the same 8 bytes of field data except if the Section Syntax Indicator has been set to zero ("0"). Private Sections may optionally use a syntax that does not follow the common format. The diagram below illustrates the content of those 8 bytes if the Section Syntax Indicator is set to "1". Only the values differ and the format and information carried following those 8 bytes is unique to the table type.

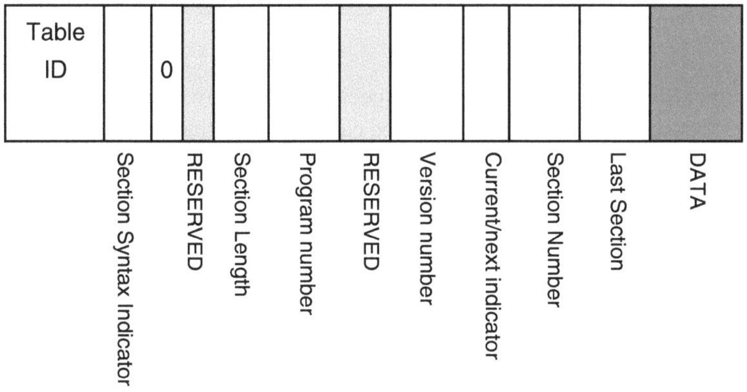

Table ID
Section Syntax Indicator
0
RESERVED
Section Length
Program number
RESERVED
Version number
Current/next indicator
Section Number
Last Section
DATA

The Program Association Table (PAT) always has a PID value of 0x00. The other information carried in the table includes its length, a 2 byte ID set by the broadcaster that uniquely identifies this TS stream from other TS streams. There is a version number that is incremented each time the PAT is updated and a bit that is used to indicate whether this is a current or future PAT.

PATs are not restricted to the 1,024 bytes allowed by the 10 bit length field. Section numbers are provided for PATs so they can be more than 1,024 bytes in total length and they also allow for acquisition of tables out of sequence. The section number enables platforms to reassemble these tables in the correct order. A last section field is provided as information to determine if an entire table has been assembled by the receiver.

As stated earlier, broadcast channels are susceptible to errors. PSI tables need to be error free so a CRC field is provided to check for delivery errors and because these tables are sent repeatedly a section that contained errors can be replaced. PSI tables can impact the bit rate that can be utilized for programming, but they also impact the time it takes to acquire programming. Network operators give this careful consideration.

The Conditional Access table has a fixed PID of 0x01 and the same fields defined for the first 8 bytes. However, the remaining bytes contain a descriptor whose content is not defined by MPEG. Descriptors, like PSI tables have common header fields but each carries data unique to its definition. Descriptors will be addressed briefly in a later section.

The Transport Stream Program Map table is a cross reference table between program numbers defined in the PAT and program elements as defined in this table. PIDs values for this table are not fixed and are selected by the network operator. Multiple instances of these tables are transmitted so that the entire content of the multiplex can be understood. A careful reading of the MPEG-2 specification is required to fully understand how to use these tables.

The Private Section table allows broadcasters to transmit information not available to all receivers but at the same time, like any private data in a Transport Stream multiplex, this information will not cause those receivers to crash. Receivers that do not understand what is contained in these sections simply skip over them when decoding streams.

9.4.4 Program Streams

As stated earlier, the MPEG-2 systems specification has two formats for delivering content. Transport Streams were described earlier. Recall that Transport Streams can carry the information associated with multiple programs while Program Streams can only contain one program. Programs Streams are primarily intended for use in stored media applications such as DVD; however they are not restricted to stored media applications. Like Transport Streams, Program Streams have a system target decoder (P-STD) that is a mathematical definition for the arrival and distribution of bytes in the stream. Both the T-STD and the P-STD models are intended to provide guidance for the correct implementation of their respective demultiplexor platforms.

Unlike Transport Streams that are used for continuous broadcasting of content, Program Streams are defined to have a beginning and end. As such, a program end code is specified by MPEG (0 × 000001B9). The hierarchy of Program Streams differs from that of Transport Streams. The pack layer is used to carry the System Clock Reference data as well as other information about the PES packets that follow, such as the multiplex rate (mux rate). Unlike Transport Streams, Program Stream uses start codes to signal the beginning of a set of data. This is a legacy of the MPEG-1 methods for delivering information to the demultiplexor.

Packs carry system headers that further describe the content carried in the PES packets that follow. Recall that PES packets in Program Streams are specified identically to those carried in Transport Streams.

Similar to MPEG-2 Transport Stream, MPEG specifies so-called Program Stream Map (PSM) tables for program streams. PSM tables are carried in a PES packet in Program Streams. This enables program and description information similar to that provided in Transport Stream PSI data to be carried in the stream. The PSM has a unique stream ID to allow it to be distinguished from PES packets that carry compressed data for both Transport Stream and Program Stream. Note, that if a PSM appears in a Transport Stream its content cannot be altered. The PSM data starts out looking like a PES packet but then the fields that follow look more like the PSI data carried in Transport Streams.

Program Streams carry a Program Stream Directory. Like PSMs, PSDs start out looking like PES packets but then differ in syntax in order to provide the information they carry. It is best that the specification be referenced in order to gain an understanding of the syntax and use of both PSMs and PSDs.

9.4.5 Descriptors

The descriptors are used to extend the definition of programs or elements within programs. Like most of the MPEG-2 systems formats these descriptors have a common syntax in their initial bytes, in this case the first 2 bytes. The first byte identifies the descriptor and the second provides the length. MPEG reserved the first 64 descriptor values and has already used many of them but has left the remaining values (64–255) to be defined and used by content providers and network operators. The syntax of the bytes following the length field of the descriptors defined by MPEG is also defined but the bytes following the user defined descriptors is not defined by MPEG.

An example of a descriptor defined by MPEG is the Video Stream Descriptor. The descriptor tag in the first byte has a value of two (0x02). The first byte following the length field indicates aspects of the video stream such as frame rate. The descriptor will also specify if there are multiple frame rates as would be the case when a movie at 24fps has commercials inserted at 30 fps. Other video information is also described. The reader should consult the MPEG-2 systems specification for the description and use of the descriptors defined by MPEG.

9.5 Summary

The MPEG-2 systems specification enables a wide range of possibilities for the carriage of bit streams. At the time when the specification was being developed, a significant portion of the specification was required to be implemented in hard wired logic. This is certainly not the case today when it is possible to use general purpose computing platforms to handle MPEG-2 bit streams as well as the new CODEC available to consumers. The only aspect of MPEG-2 systems that is still recommended to be implemented in hardware specific circuits is PCR extraction and SCR frequency control. If the remainder is implemented using software or microcode, care should be taken to understand expected bit rates that the demultiplexor will see, as well as the bit rates of individual video, audio or even data stream in order to ensure smooth execution.

The MPEG-2 Systems specification contains informative annexes that provide examples and are intended to assist implementers with a less formal description than is contained in the normative sections of the specification.

Not all of the capabilities of the current MPEG-2 Systems specification have been described in this chapter. MPEG-2 Systems has continued to evolve and incorporate the ability to define and carry the data associated with the additions to the MPEG committee's suite of technologies. This includes new audio and video technology as well as metadata. This is a testimony to the foresight and flexibility built into the original specification that was first completed in November of 1994.

Chapter 10
MPEG Multimedia Scene Representation

Young-Kwon Lim, Cyril Concolato, Jean Le Feuvre, and Kyuheon Kim

10.1 What is MPEG Richmedia?

10.1.1 Introduction about Richmedia

Richmedia is a recent term used to identify a media beyond conventional multimedia. The term multimedia is used to identify a media as a temporarily synchronized combination of more than one media elements such as audio, video, text. Richmedia extends the scope and the capability of multimedia by introducing graphics as an additional element. Richmedia also adds new features such as spatial composition of elements and interaction among the elements of the media and the user. Therefore, Richmedia is defined as a combination of more than one media components such as audio, video, text and graphics with the support of spatio-temporal composition of components and the interactivity among the components and the user.

In comparison to multimedia, capability of spatio-temporal composition of richmedia provides a way to construct the media as a combination of multiple audio-visual media component in time and space. More specifically, functionality of temporal composition enables richmedia to be constructed to present more than one video component with its associated audio component at some point in time. This feature introduces the concept of "scene". MPEG richmedia standards provide technologies to digitally represent and manipulate the scene.

MPEG has been working on the standards on richmedia since the beginning of the MPEG-4 standard and produced two major standards as a family of MPEG-4, Binary Formation for Scene (BIFS) and Lightweight Scene Representation (LASeR.) Following chapters will provide detailed technical descriptions about each technology.

Y.-K. Lim (✉)
Media Innovation Center, Goyang City, Gyeonngi-Do, Korea
e-mail: yklim.korea@gmail.com

L. Chiariglione (ed.), *The MPEG Representation of Digital Media*,
DOI 10.1007/978-1-4419-6184-6_10,

10.1.2 Technical Components of MPEG Richmedia

MPEG richmedia standards share several requirements. Here are major technical components included in MPEG richmedia standards.

- Temporal composition: Richmedia standards provide means to specify the order of components in time. Components can be placed in time relative to each other or have absolute temporal distance from the beginning of the presentation. Richmedia standards also support synchronization between components by sharing same clock reference.
- Spatial composition: Richmedia standards provide means to place a component in a specific position of 2D or 3D space based on its own coordinate system. In 2D spatial composition, it can specify the rendering order of components in addition to the position to realize layering concept. Richmedia standards also provide means to specify the size and the shape of component to be rendered on the scene. Position and size can be specified either as an absolute value or as a relative value. All spatial composition information is not static so that they can be dynamically changed during presentation by using an update mechanism.
- Interactivity: Richmedia standards allow the user to modify spatio-temporal composition information of components during the presentation. Interaction between the user and a certain component of the richmedia scene can be propagated to the components in the same presentation. In other words, one interaction from the user can consequently result in multiple changes of other components. Conditional interaction results in different changes according to the status of the component when the interaction in place is also supported.
- Vector graphics and animation: Since graphics components play a key role in distinguishing richmedia from conventional multimedia by enabling richer presentation than its predecessor, a richmedia standard provides its own means to represent 2D or 3D vector graphics components tightly coupled with its spatial composition mechanism and conceptual rendering system. By dynamically modifying the parameters related to vector graphics component, richmedia standards also support animation of presentation.
- Text and fonts: Text is another component whose importance is much higher than the case it is used as part of multimedia. Various rich rendering features for text are supported by richmedia standard including means to specify the fonts to be used as part of the standard itself or by a referenced standard.
- Update mechanism: Richmedia is time-based content consumed as a sequence of audio-visual representations at each specific time. Therefore, richmedia standards provide means to segment the presentation information into a number of instances for modification at certain time during the presentation. This feature enables to deliver richmedia presentation as a combination of static self-contained representations of a certain time and the series of subsequent updates modifying specific part of the representation in a manner that evokes I frames and P frames of video coding.

- Binary representation: Unlike other richmedia standards or propriety technologies, richmedia standards developed by MPEG provide a built-in mechanism to represent the information in binary format. Binarization of presentation information brings reduction of the size of the data to be delivered. In addition, it brings computational processing efficiency at the hardware level by minimizing the number of clock cycles required to understand the command and the value of its instruction. Indeed the processing of text based representation requires the step of identifying the keywords delimited by the white space characters, which is not required for the processing of binary representation.

10.1.3 Future of Richmedia Technology

As powerful content consumption devices with high resolution, large screen and fully interactive broadband network connection get popular, the role of richmedia in the media services becomes more important. Unlike legacy multimedia service such as analogue or digital broadcasting, in the emerging media services such as IPTV, ConnectedTV, and SmartTV the user is given a high number of choices as a combination of many media assets. Moreover services are mostly initiated by user selection and consumption is fully controlled by the user during the presentation. In other words, in the emerging media services, an important requirement is the possibility for the service provider to rely on user interfaces specifically designed for each content so that the service can be distinguished from other services by highlighting its major characteristics and giving the user available options for consumption. Therefore, in the emerging media services it is very important to provide means to present dynamic descriptions about the content to the user as a combination of various components and to provide full controllability to the user by allowing interactive re-composition of media and selection of specific components and spatio-temporal fragments of them for consumption matched to the user. This will accelerate the market adoption of richmedia technology. In addition, increasing market demand on the capability of one source multi use of content will highlight its power of richmedia standard ensuring cross platform interoperability.

One major architectural challenge related to MPEG richmedia standards is providing a framework for spatio-temporal composition without complex vector graphics component. Even though BIFS defines scene graph profiles to include technologies for scene representation without including technologies specific to vector graphics, the basic framework of BIFS is still quite complicated because it needs to support both scene representation technologies and vector graphics technologies. So, it would be worth considering for the coming standard to design an architecture based on fully separable platform for spatio-temporal composition technologies and vector graphics technologies.

10.2 MPEG-4 BIFS

10.2.1 Overview of the Standard

Binary Format for Scenes (BIFS), is defined in part 11 of the MPEG-4 Standard [1]. BIFS was primarily developed with interactive TV in mind, but it actually covers a wide range of applications, from mobile TV to 3D gaming on desktop or set-top boxes. BIFS provides a standard language to create rich, interactive multimedia content, and a binary coding to efficiently transmit this content. BIFS extends the Virtual Reality Modeling Language (VRML) [2] in various aspects such as 2D graphics, text, 3D Avatars, advanced media playback or user interaction tools. The 2D features of the standard are equivalent to common tools found on the World Wide Web, such as HTML or SVG, while the 3D part of the standard offers similar features as 3D modeling languages such as Collada [3] or X3D [4].

BIFS follows MPEG's Rich Media approach. The content, called a scene, is delivered to the client using data streams. The server can modify the scene at any time by sending scene updates to the client. These updates allow insertion of new content, modification of existing properties such as color or text, or modification of the existing interactivity logic.

The strength of BIFS is its ability to describe interactive presentations with mixed 2D and 3D graphics under a unified model. Scenes are expressed in a compressed and memory efficient way, even in constrained environments. BIFS scenes can be presented on most multimedia delivery architectures such as Digital TV (MPEG-2 TS) or IP environments (RTP or HTTP streaming). The standard is quite simple to understand and does not require the author to understand the complex set of recommendations from XML/W3C to design appealing applications.

At the time of writing of this book, the main weakness of the standard is the lack of powerful authoring tools for complex interactive applications. Unfortunately this is often the case for most open scene composition formats. Commercial authoring tools and players are available, and readers may also want to check existing open implementation of the standard, the MPEG-4 Reference Software [5] freely available from ISO or GPAC [6], an open-source MPEG-4 implementation licensed under LGPL.

10.2.2 Structure of a BIFS Scene

The BIFS language is structured around the concept of nodes, which are used to represent specific parts of the content. Some nodes can be used to group together other nodes, and therefore define a hierarchy of nodes. Groups of nodes can be added, deleted or modified by the server or by user interactions.

Two types of nodes are particularly interesting: leaf nodes and grouping nodes. Leaf nodes typically refer to graphical primitives such as 2D curves or 3D meshes,

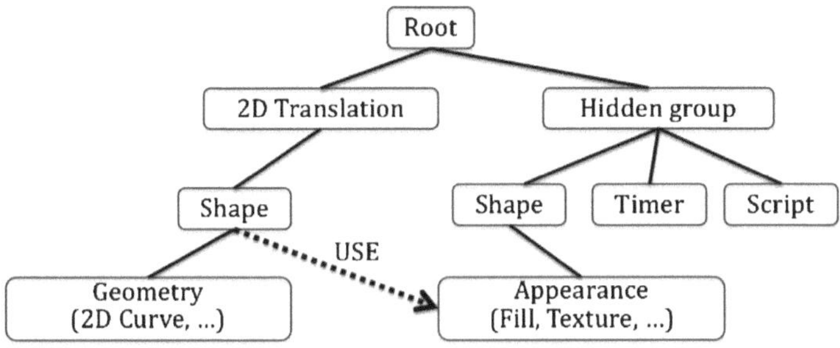

Fig. 10.1 Logical example of a scene tree

texturing or painting information, or interactivity-related data. The BIFS scene is made of a single top-level node grouping all other nodes used in the content; this hierarchical set of nodes is referred to as the scene tree or scene graph.

A given node will not necessarily fit in any parent; for example, a *Rectangle* node can only be a child of a *Shape* node, which in turn can only be a child of a 2D or 3D grouping node. From this restriction, a classification of nodes called Node Data Type is derived, identifying for example 2D nodes, top-level nodes, texture nodes, geometry nodes or all nodes.

Nodes can be identified within a scene by assigning them a unique identifier, which can be an integer or a string, depending on the BIFS encoding configuration. This allows a server to send a command targeted at a given identified node. Within one scene tree, there may not be different nodes with the same identifier.

In order to simplify content design, it is possible to reuse a node in various places of the scene tree, as shown in Fig. 10.1. One example could be the reuse of a rectangle with a different text on top of it when creating buttons. However, cyclic references of nodes such as A->B->A are not allowed in the standard.

Each node in a BIFS scene is described by one or several properties, called fields. Fields have predefined types among Boolean, integer, float, double, RGB color, 2- 3- and 4-dimensional vector, string, uncompressed image (pixels arrays) and node. Additionally, each field has an associated access method: a field can be read-only or read-write. Some nodes may trigger events in the scene, through mechanisms explained in the next section, in which case the event will be available in a field with a specialized read-only access method called *eventOut*. Some nodes may also use events generated by other nodes for interactivity purposes, in which case the event will be received in a specialized write-only field called *eventIn*.

The connection between *eventOut* and *eventIn* is called a *ROUTE*, and it simply instructs the client to copy the value from the *eventOut* to the *eventIn*. This implies that *ROUTE* can only connect fields of the same type. When different types need to be connected, BIFS provides a specific node, called *Valuator*, which allows converting from one type to another by adding or suppressing dimensions from input events, as shown in Fig. 10.2.

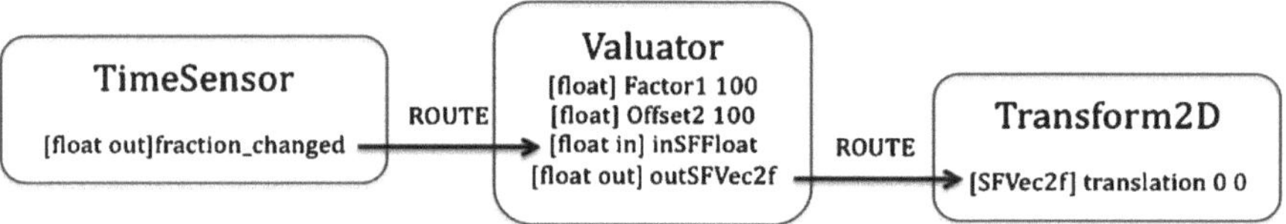

Fig. 10.2 Example of valuator converting a fractional value in [0,1] to a 2D vector value in [{0,100}, {100,100}]

ROUTEs can have identifiers attached to them, to allow deletion and modification of the *ROUTE* in order to modify the interactivity logic. Fields with read-write access are called *exposedField* and can be the source or target of *ROUTE* to or from other nodes.

Another key feature of BIFS is the ability to allow an author to define new nodes not described in the standard. While, obviously, new functionality cannot be added, this tool allows creating higher level nodes, for example a "Button" node can be created to replace a complex group involving a *Rectangle node*, *Text node* and mouse or key sensors. The definition of such new nodes is called a *PROTOTYPE*. This definition describes the new node syntax, its properties (fields) and its implementation using existing standard nodes or other *PROTOTYPE* nodes. The *PROTOTYPE* node can later be used to create a new node in the tree as if the node was a predefined one, and all the BIFS tools (events, *ROUTE*, node re-use…) can be used with this node. This tool can be used to define libraries of components that can be reused in other contents by simply giving the location of the library.

Content may originate from different locations in a BIFS scene. There are various ways of referring to external content for images, video or audio data using Uniform Resource Locators (URL). Most importantly, the content can be split in sub-presentation independently playable and referred to in the main content through the Inline node. This simplifies content design and, most importantly, allows reusing existing pieces of content when designing interactive portals.

10.2.3 Composing Media Objects

BIFS describes objects to be drawn on the screen with a specific node called *Shape*, with the exception of the *Background* or *Background2D* nodes, which draw a background or a sky dome using color or images. *Shape* nodes can be used for drawing 2D or 3D graphics, for text, movies or images. The different properties of the *Shape* node describe specific aspects of the drawing. This includes the appearance and geometry of the shape.

Possible geometries supported in BIFS include text, high-level shapes such as rectangles, circles, polygons, curves, boxes, cones, spheres or cylinders and low-level shapes such as 2D curves, elevation grids, extrusions or more elaborated 3D meshes.

For example, complex geometries can be described using the *XCurve2D* node, which allows combining lines, quadratic or cubic Bezier curves and elliptical arcs. A special node called *Bitmap* can be used to draw video and images at their native resolution.

The appearance associated with the shape allows describing the visual style of 2D and 3D objects. The standard provides a large choice of styles: image and video texturing on shapes (including text), solid color with alpha-blending, linear and gradients patterns with opacity, color matrices modifying the colors of all synthetic objects in a sub tree, texture transformations such as blending, blurring, color keying. The standard allows for fine outline control, including line caps (termination) and join, dashing, opacity and texturing with image, video or gradient.

Like most graphics languages, BIFS uses absolute positioning of graphics: each grouping node defines its own local coordinate system, and most geometric primitives in the standard (*Bitmap*, *Rectangle*, *Circle*, *Box*, *Sphere*...) have their center co-located with the origin of the local coordinate system of their parent node. One noticeable exception to this is the Text node, where by default the first glyph of the string is anchored at the local coordinate system origin. As for the scene as a whole, the origin (x = 0, y = 0) of the root coordinate system is at the center of the drawing area, with the y-axis in the up direction and, for 3D cases, with the Z-axis oriented towards the user. This differs from most 2D graphics system where the origin is at the top-left of the drawing area and the y-axis in the down direction. The reason for this choice is to unify 2D and 3D coordinate systems, as BIFS can handle both types of graphics in a single content. This unification also explains why BIFS units can be pixels, mostly used in 2D graphics design, or meters, default unit in most 3D modeling tools. Some grouping nodes, such as *Transform2D* or *Transform*, may be used to modify the transformation between the parent and the local coordinate system, by specifying translation, scaling or rotation parameters in 2D or 3D. By default when drawing 2D primitives, BIFS will draw each node in the order of appearance in the tree; this is commonly referred to as the "Painter's Algorithm". This behavior can be changed using the *OrderedGroup* node, which allows modifying the order in which its children nodes are rendered. The *Switch* node can also be used to render one or none of its children rather than the entire sub tree. As absolute positioning is not always the best choice when designing complex interactive applications, BIFS also provides tools for relative positioning, or "flow layout," in which case the position of a graphical primitive depends on the position of the previous one. For example, the *Layout* node performs automatic horizontal or vertical rectangular layout of text and graphics, with optional wrapping and scrolling; or the *PathLayout* node performs linear layout along any given curve. The language also defines tools for 2D graphics clipping by 2D shapes, and hierarchical clippers can be combined in union or intersection mode.

As previously stated, BIFS is able to handle both 2D and 3D content. There is however one restriction to consider: 3D content can be rendered in 2D content only when a parent node of type *Layer3D* or *CompositeTexture3D* is defined, which will define a 2D rectangular area where 3D content can be rendered. This restriction

Fig. 10.3 Example scene of mixed 2D and 3D

simplifies hardware integration on most devices, where 3D rendering is often a dedicated component of the graphics engine, not integrated with 2D rendering.

Figure 10.3 shows a mixed 2D and 3D scene displaying a single 3D model from different views using *Layer3D*. The model is inserted in the scene using an *Inline* node, which is then re-used three times with different lighting, backgrounds and camera configurations. This example also shows the layering of textual 2D data with 3D graphics.

The drawing of nodes is usually performed on-screen, but it is possible to draw part of the scene to some off-screen memory and use the result graphics as a source image for texturing an object; this is achieved with *CompositeTexture2D* and *CompositeTexture3D* nodes. The textures defined by these nodes can use any of the language features, in particular animations and pointing device interactivity, which is particularly useful when integrating 2D user interfaces in 3D worlds, as shown in Fig. 10.4.

BIFS also provides sophisticated management of audio data in the presentation. The language allows physical modeling of a virtual world by specifying reverberation, reflectivity and transmission properties of the objects in the scene (*AcousticMaterial* or *AcousticScene* nodes); each sound source can be spatialized (*Sound*, *Sound2D*, *DirectiveSound*, *SurroundingSound*, *WideSound* nodes), spatialization can be modified (*Transform3DAudio* node) and the listening position of the viewer can be specified independently from the 3D viewing position (*ListeningPoint* node), in order to use BIFS audio tools independently from the presence of a 3D graphic context. Various post-production tools can be added to the content, such as delay (*AudioDelay* node), mixing, chorus, compressor, echo or flange (*AudioFX* node). The standard also provides means, through the *AudioBuffer* or *Advanced AudioBuffer* nodes, to store short audio clips in memory and replay them at will, as is often the case in user interfaces. As they are defined through nodes, all these tools can be interactively configured during the presentation.

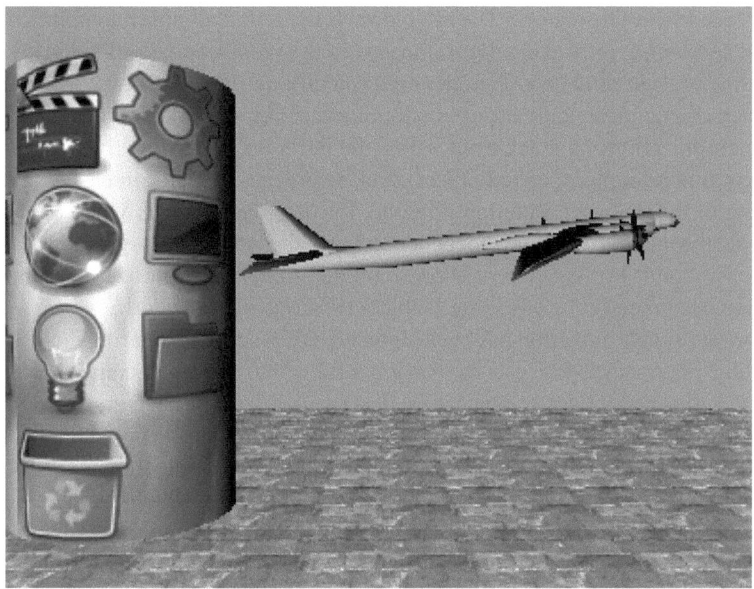

Fig. 10.4 Example of 2D User Interface integrated in 3D scene

10.2.4 Animating Objects

Rich media presentations are intrinsically dynamic: objects are animated in position, color or transparency. Animations may be dependent on a video timeline, and media can be muted, paused or restarted... The BIFS standard provides two possibilities to build animations.

The first possibility is the use of timers, through the *TimeSensor* node, which emits events, some of them once per drawn frame, indicating the current state of the timer such as activation/deactivation or current fraction of elapsed time since the start of the timer. These events can then be sent to other nodes to animate the scene. A typical usage of the time sensor is to route its output to interpolator nodes. These nodes use author-defined fraction/value pairs, take a fraction as an input event, find the closest fraction values defined in the node and use the value pairs corresponding to interpolate the output value. The interpolation can be linear (*Interpolator* nodes in BIFS terminology) or follow a more complex path such quadratic or cubic Bezier (*Animator* nodes). *Interpolators* and *Animators* exist for most predefined types of the language: color, floats, 2D and 3D vectors, Rotation.

The second possibility offered by the standard is to use scene updates to modify the content. These updates can be in the stream carrying the scene, or in a dedicated stream, whose playback can be controlled in the scene through the *AnimationStream* node. The advantage of this approach is that complex animation operations can be performed, including node removal and insertion, which cannot be achieved by interpolators. The main drawback is that all possible points in time are not addressed: increasing the frame rate of the presentation will not bring more smoothness. *AnimationStream* nodes are typically used to control the synchronous playback of graphical annotations or subtitles describing a video or an audio object.

Synchronization is not achieved within the BIFS language directly. It is achieved through time references, called object clock references, which can enslave a media timeline to the main presentation timeline, for use cases such as video annotation, or make the media timeline independent from the scene one, typically used for interactive control of a video or short, looping animations. These synchronization rules are described in the associated Object Descriptor Framework.

If proper synchronization has been signaled, BIFS supports altering media speed, including play/pause/stop use cases, or seeking into media streams through the *MediaControl* node. Each media can be described as a set of chapters, called Media Segments, which can be used to control the media playback or monitor chapter changes. The *MediaSensor* node allows an author to determine the media duration, current playback position in seconds and current chapters, which can then be used to synchronize animations in the scene with the media. It is important to notice that media management tools in BIFS are independent from the media type and can be used with audio, video, streaming text or other animations.

Finally, BIFS addresses the complex issue of synchronization with a media delivered on a jittery, unreliable network such as TCP/IP. The standard provides ways, through the *TemporalGroup* and *TemporalTransform* nodes, to automatically stretch or shrink the timeline of the scene to follow the timeline of a given media, ensuring proper synchronization is performed between animations and media, independently of playback stalling due to network conditions.

10.2.5 Interacting with the Content

Interactivity is one of not the most important features of Rich Media. It enhances the user experience (customization, bonus tracks, voting…). Interactivity in BIFS follows a classic approach used in Rich Media. Events, or notifications, are triggered by different sources and are propagated in the presentation to perform various actions potentially triggering new events.

In BIFS, events are initially triggered by specific nodes called sensors. Events can be of many types: pointing device clicks and moves (*TouchSensor* node), pointing device drags for translation (*PlaneSensor2D*, *PlaneSensor* nodes) or rotation (*DiscSensor*, *CylinderSensor*, *SphereSensor* nodes) in 2D or 3D space, navigation keypads (*KeyNavigator* node), scene time events (*Timer* node), media time events (*MediaSensor* node); a special node, *InputSensor*, allows interfacing with the scene any kind of interaction devices: keyboard, remote controller, accelerometer…

Once an event is triggered, it can be routed to any number of nodes in the scene for further processing. Each event processed may imply generation of new events, which will be queued, and processed once the triggering event has been sent to all destination nodes. In order to avoid recursive cases, a player insures that each route is followed once per frame.

Event processing can take several forms in BIFS. The simplest form consists of passing event values to target nodes, as seen in the interpolation examples, potentially modifying them through a *Valuator* node. The second form, unique to MPEG Rich Media, is to execute a set of scene updates when receiving the event. This is achieved through the *Conditional* node. This node holds a set of BIFS scene updates to execute when activated. Any BIFS scene update can be used in this node, allowing node deletion or modification or even route insertion and replacement. This functionality of BIFS allows content creator to design complex interactive scenario even in the absence of scripting, which may be the case in constrained devices. However, this functionality cannot address complex cases where mathematical operations are needed. The third form of event processing relies on scripting. In this approach, the event is routed to a *Script* node. This node will execute some code depending on the event value. Script code in BIFS is typically described using the VRMLScript language. This language is a variant of ECMAScript language [7] (like JavaScript) but with various specific extensions for handling BIFS predefined types such as nodes or vectors. For the most advanced cases, BIFS also defines *MPEG-J*, a framework where a BIFS player is coupled with a Java virtual machine, which can execute Java code to manipulate the various subsystems involved in the BIFS scene, such as scene graph, network stacks, media decoders, system resources and capabilities and MPEG-2 Transport Stream filtering. Java applets, called *MPEGLet*, are carried in dedicated streams and use their own decoder, whose task is to decide when an *MPEGLet* is ready to be loaded in the virtual machine and ready to be started. The Java code can then register even listeners on nodes of interest and be notified of events when they are triggered.

Interactivity usually implies upstream communication with a remote server, to query specific information to populate the content or post a user choice (Video On Demand menu, voting…). BIFS provides a specialized node, *ServerCommand*, which allows sending back information to the originating server. The server may in reply send modifications to the scene using the delivery channel of the BIFS content. This approach slightly differs from the AJAX model where the reply is sent using the delivery channel of the request.

The *Anchor* node provides another approach for client/server interactivity. This node allows navigating between different presentations. When the URL of the *Anchor* node is built with CGI parameters, each requests for a new content instructs the server to customize the scene. This approach is common in many web portals using PHP or JSP.

When navigating between scenes in an interactive portal, it may be useful to be able to locally save and restore some information about the various scenes, using a mechanism similar to HTTP cookies. In BIFS this is achieved with the *Storage* node. This node provides local storage of data regardless of the transport protocol used: online or offline, broadband or broadcast with no back channel.

Finally, the interactive toolset of the BIFS language can be used to dynamically adapt the content to the terminal characteristics or user preferences; the *TermCap* and *EnvironmentTest* nodes allow querying simple terminal characteristics (width, height, pixel screen density, presence of touch screen or keyboard...) while the *ReplaceFromExternalData* command enables complex scenario such as querying a value from a device database such as a user profile.

10.2.6 Compressing and Exchanging BIFS Content

As stated previously, BIFS is a binary format. The binary encoding of the scene tree is a lossless and contextual process. Each node is declared through a binary tag dependent from where the node can be used. For example, a node will have a different tag when coded as a child of a 2D grouping node or of a 3D grouping node. This allows reducing the number of bits required to identify the node.

Node fields are identified in the bitstream using similar techniques. When coding a node, fields corresponding to events (*eventIn* or *eventOut*) are not coded since they do not carry any value. This reduces the number of bits required to identify a given field.

As opposed to the structure of the scene tree, the values of the node fields can be encoded either in a lossless manner or in a lossy manner, when quantization is used. Quantization in BIFS is a linear quantization process, where floating-point values are clamped to a given interval $[V_{min}, V_{max}]$ and coded in this interval using a given number of bits *NbBits*. The inverse quantization of a value V_q read from the bitstream is

$$\hat{v} = V_{min} + V_q \frac{V_{max} - V_{min}}{2^{NbBits} - 1}$$

Specialized quantization methods exist for integers, rotations and normal vectors but they use the same principle. V_{min}, V_{max} and *NbBits* can be specified for each predefined data type of the language (e.g., 2D vector, float, color, ...), through the use of a *QuantizationParameter* node in the tree. The quantization then applies to the subtree of this node's parent, and is no longer active once the BIFS command has been decoded. It is also possible to define a global *QuantizationParameter*, which enables quantization on the entire scene tree for all decoded commands.

More advanced tools are available for compressing specific data types in the BIFS scene. We can cite predictive coding for arrays of values. In this case, values in the array are coded as differences to previously decoded values, in an approach similar to predictive coding in video. The BIFS Animation (aka BIFS-Anim) is another coding tool of the standard. It codes, in a dedicated stream, the modification over time of a set of values in an array using the same predictive technique. The set of values is defined in the configuration of the BIFS-Anim decoder.

BIFS not only needs to be compressed, it first needs to be authored. In order for authors and authoring tools to exchange their content in a lossless, human-readable format for debugging or modification purposes, an XML-based representation of the scene has been defined: XMT, the Extensible MPEG-4 Textual format. This format defines two languages: XMT-A, inheriting from X3D, which provides an exact XML representation of the binary format construction, which makes it particularly suited for use cases where fine control over the encoder is required; and XMT-Ω, extending the W3C SMIL language, which provides higher level tools to describe a scene but does not uniquely translate to XMT-A or BIFS. Both languages allow representing the multimedia content over time by using *par* containers encapsulating the scene updates that will be received by the MPEG-4 terminal.

10.2.7 Delivering a BIFS Scene

As discussed previously, a typical BIFS scene may involve composition of several media (audio, video, subtitles) for final display, and some specialized streams such as BIFS-Anim or MPEG-J streams may be present; obviously, different decoders will be needed to handle these various media types, which need to be signaled in the content. Part 1 (Systems) of the MPEG-4 standard [8] defines a set of tools, the *Object Descriptor Framework*, in order to describe the different streams used in the scene, their associated decoders, their dependencies in time or decoding order as well as various meta-data such as language, alternate bit rate versions of the same stream, quality of service management... The core notions of this framework are the *ElementaryStreamDescriptor* (*ESD*) and the *ObjectDescriptor* (*OD*). The *ESD* is a set of information describing an elementary stream, such as an audio or a video stream, the synchronization rules that govern transport and decoding of this elementary stream, *IPMP* (Intellectual Property Management and Protection) rules for protected streams, and various meta-data. It is important to notice that BIFS data itself is carried in an elementary stream. Each elementary stream is assigned a type indicating the nature of the stream: audio, visual (video and images), text, MPEG-J, BIFS...

The *OD* is a set of *ESD* used to link encoded media with the scene. Each *OD* is assigned an identifier called *ObjectDescriptorID* and a BIFS node such as *MovieTexture* or *AudioSource* may refer to an *OD* through a specific URL scheme in the form of ***od:ObjectDescriptorID***. The framework also takes into account that different objects may appear or be removed during the lifetime of a presentation; in a design similar to BIFS, *ODs* are inserted or removed through commands delivered in a dedicated elementary stream. An *OD* can be a collection of several streams:

- With the same type: this describes either streams representing alternate versions (bit rate, language, resolution, coding format...) of the same content if the streams do not depend on each other for decoding, or hierarchical scalable streams otherwise.
- With different types: it is only allowed to have one extra meta-data stream, one extra clock reference stream or one extra IPMP stream to an *OD* describing a media. This means that a single *OD* cannot describe both an audio stream and a video stream. An exception to this rule is that streams carrying *OD* and streams carrying BIFS can be aggregated in a single object.

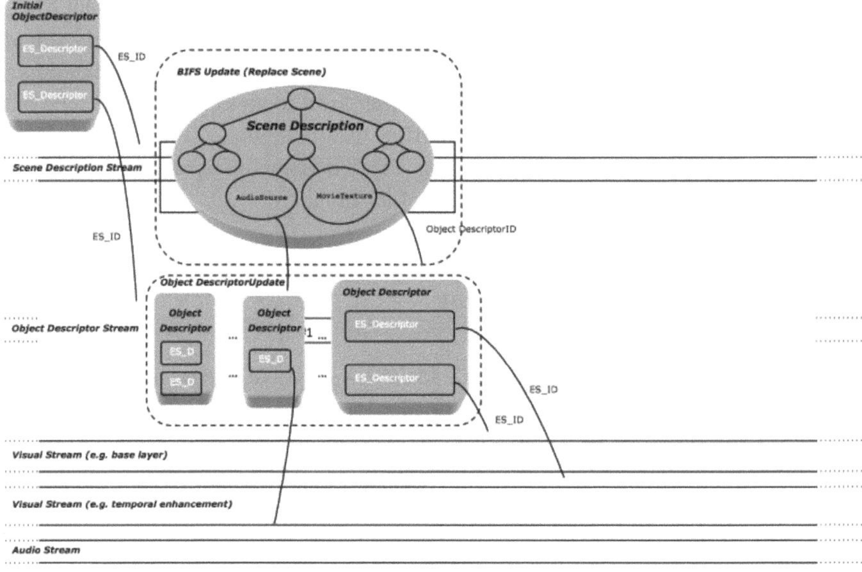

Fig. 10.5 Content access procedure

In order for an MPEG-4 terminal to start the presentation, an initial *OD* needs to be acquired: the *InitialObjectDescriptor* (*IOD*). An *IOD* usually contains *ESD* for *BIFS* and *OD* streams, and indicates the profile and levels needed to process the content for each media type (audio, video, text, BIFS, OD…).

The delivery of the *IOD* depends on the underlying transport protocol: in the Program Map Table when the content is carried over MPEG-2 Transport Stream or in an SDP file when carried over RTP The terminal then opens the streams described in the *IOD* and starts processing BIFS and OD data as shown in Fig. 10.5.

10.2.8 Profiles and Levels

As with most MPEG technologies, BIFS uses profiles and levels to select a subset of the tools defined in the standard for commercial deployment. BIFS comes with two orthogonal profile families: the *Graphics* profiles, which cover graphic tools such as text, curves, lights, meshes, texturing; and the *Scene Graph* profile, which cover all other scene constructs such as sensor nodes, grouping nodes, script or scene commands. To completely define any MPEG profile, levels are needed to exactly define the minimal requirements of a decoder. Levels on the *Graphics* profile specify the degree of complexity for the graphical elements: maximum supported number of nodes of a given type, maximum supported number of points in polygons, maximum number of text characters in a scene… Levels on the *Scene*

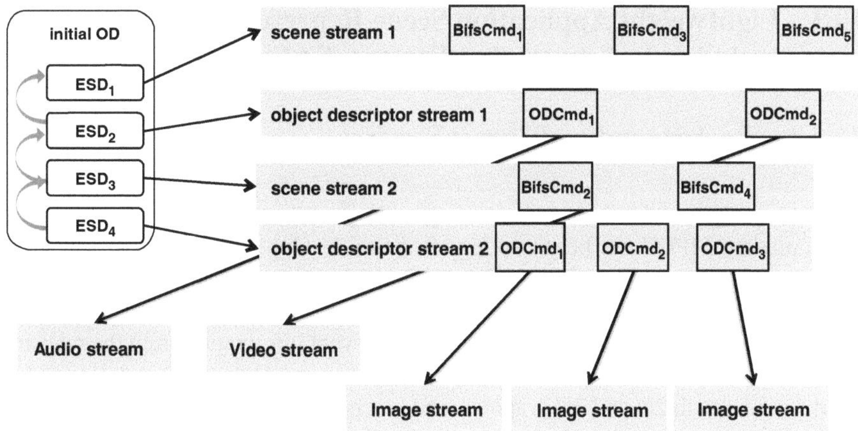

Fig. 10.6 Example of scalable scene for T-DMB

Graph profile specify the degree of complexity of the scene structure: maximum number of nodes, maximum number of children per given grouping node, maximum number of key/value pairs in interpolators, fields to be ignored by the decoder. Combined together, *Scene Graph* and *Graphics* profiles range from *Basic2D*, which only allows for audio-visual presentation with no interaction or graphics, to *Complete*, supporting the complete set of tools defined in the language.

10.2.9 Example of BIFS Commercial Deployment: T-DMB

The BIFS standard has been adopted by the T-DMB specification [9], used in Korea and other countries for terrestrial broadcasting of digital multimedia content to mobile devices such as phones or car entertainment systems. The T-DMB standard uses the *Core2D* profile of the BIFS specification, therefore enabling simple interactivity and graphics to be embedded with the audio-visual content. The profile uses scene commands for interactivity rather than scripting, thereby keeping the complexity and CPU requirements low, which make it particularly well suited for low-complexity portable devices. In order to support devices without interactivity support, the BIFS and OD streams in the T-DMB broadcast are organized as a scalable presentation, as illustrated in Fig. 10.6.

The *IOD* describes two dependent sets of BIFS and OD streams. The first set only carries the base presentation, which consists of a simple BIFS scene with at most one audio and one video object, whose descriptor are carried in the OD stream of this set. The second set describes all the interactive services, which is usually carried in one or two BIFS streams and one OD stream in charge of declaring the different images used by the service. The specification restricts the presentation to at most only one audio and one video stream.

10.3 Lightweight Application Scene Representation (LASeR) and Simple Aggregation Format (SAF)

10.3.1 Background and Target Services

As the convergence between broadcasting and communication has been emerging into a market, MPEG has been making efforts to develop richmedia representation technologies. BIFS, the first Richmedia standard provides the functionality of linkage between a TV program and its relevant data, that is, when and where the relevant data is presented in the TV program, and provides the event information from the relevant data to the other one to be presented in the TV program. Also, BIFS is designed for the richmedia with 2D and 3D graphics, and is represented in terms of a rather complicated binary code for scene compositions. These properties in BIFS caused the difficulties in its usage for resource limited devices such as mobile devices.

In order to overcome these limitations, in 2004 MPEG started to develop a new standard called LASeR (Lightweight Application Scene Representation) [10], which supports only 2D graphics, and is represented in terms of both a XML based description and a simpler binary code. Thus, these properties make LASeR to be more suitable for improved and efficient richmedia applications and services such as mobile TV and web-like navigation services on mobile handsets.

The first version of the LASeR standard was finished in 2006 and later (in 2008) LASeR extended with Scene Adaptation functionalities for customized services such as one-source and multi-uses. A third extension, providing an efficient technique to present Structured Information (SI) on a scene by referencing specific portions of SI from Presentation Information (PI), has been produced in 2009 with the name LASeR PMSI (Presentation and Modification of Structured Information). This allows for dynamic updates and object-based interactivity such as widget-like rich media applications over broadcasting and telecommunication.

10.3.2 LASeR

10.3.2.1 Overview of LASeR

LASeR is a standard enabling the description of a scene of richmedia services including interactive functionality. The scene is described in two ways: text and binary. Thus, a richmedia service is specified by an XML based scene description and by its binary format for transmission over a variety of protocols as shown in figure.

As we can see from Fig. 10.7, the scene description functionality in LASeR was developed on the basis of SVG (Scalable Vector Graphics), and W3C (World Wide Web Consortium) vector graphics standard [11]. In addition, more functions such as interactivity are provided as LASeR extensions. At the initial stage of

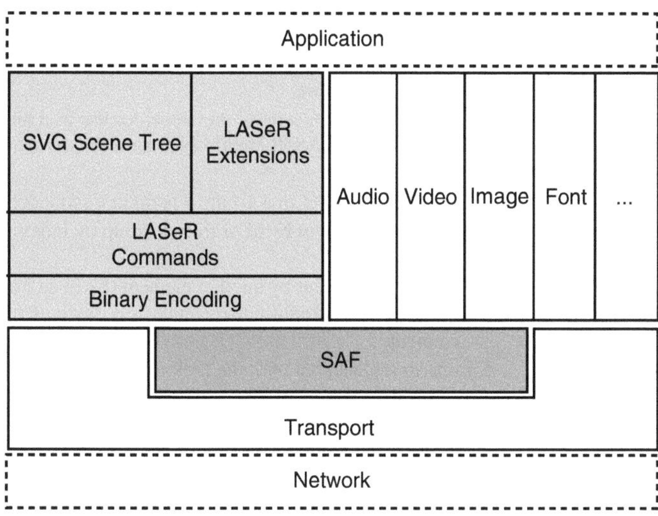

Fig. 10.7 Structure of LASeR and SAF

development of LASeR, SVG 1.1 was used as a basis for scene description, but LASeR has eventually adopted SVG Tiny 1.2 version, approved by W3C in 2008, for supporting powerful interactivity and visual applications such as device GUI, 2D Game, idle screen etc.

The scene description part explained above yields mainly three important functionalities as follows:

- Spatial and temporal relationship in a scene among the objects such as text, media, graphics, animation etc.
- Interactivity function among those objects in a scene
- Update function of a scene description when a scene has been changed

In order to more efficiently change a scene description, LASeR has brought a command function as used by MPEG-4 BIFS. The list of commands is shown in Table 10.1. For a scene change, the scene elements and commands are binary-encoded and packetized into a transmission format called "Simple Aggregation Format (SAF)" for transmission.

10.3.2.2 LASeR Scene Description

As explained in the previous section, LASeR provides spatial and temporal relationships, and interactivity functions, composed of both SVG Tiny 1.2 version and LASeR extensions.

The specification of SVG Tiny 1.2 is defined in terms of XML scheme structure. Since SVG describes a scene in terms of XML, it can be parsed and rendered by

Table 10.1 Scene command of LASeR

Commands	Description
NewScene	To create a new scene
RefreshScene	To repeat the current state of the scene, for use as a random access point into the LASeR stream or as a means to recover from packet loss
Insert	To insert any element in a group, a point in a sequence
Delete	To delete any element by id or from a group by index, a point in a sequence
Replace	To replace an element by another element (by id or from a group by index), or to replace the value of any attribute of any element
Add	Similar to replace, but with the notion of adding to the value rather than replacing it
Save, Restore and Clean	To save, reload or remove persistent scene information in the form of the value of a list of attributes. Other commands have no influence on persistent scene information
SendEvent	To send an event to any element in the scene
Activate	To transfer an element from the waiting tree to the scene tree
Deactivate	To transfer an element from the scene tree to the waiting tree
ReleaseResource	To instruct the LASeR engine that a resource (typically media stream) will no longer be used in the scene and may therefore be reclaimed

web browsers, and edited by text editors. Also, it supports hyper link and java script. The characteristic of vector graphics allows to easily scale a scene up and down without any quality degradation. As an example, the scenes with combination of animation, hyper-link, interactivity and scripting described by SVG are shown in Fig. 10.8.

LASeR extends SVG Tiny 1.2 to provide interactivity function and additional scene description function. Figure 10.9 shows an example of using one of the elements specified in LASeR extensions such as "conditional" element, where an event "ev:listener" occurs and produces ID "s1" when a "up" in a keyboard is pressed, and the "conditional" event with the same ID "s1" makes the relevant command realized, that is, the element in the group "g" is translated into the position of "-5". Thus, the "conditional" element provides the interactive functionality of enabling the LASeR commands for scene change or update.

The other elements provided by LASeR extensions are as follows:

- Conditional: to allow sets of scene commands to be inserted in the scene, for later execution upon activation by time or through the XML Events listener element.
- CursorManager: to point to any LASeR element to confer it the semantics of virtual pointer.
- RectClip: to acts as a clipping element, limiting the rendering of its children to a device-pixel-aligned rectangle, in addition to the semantics of the SVG g element.

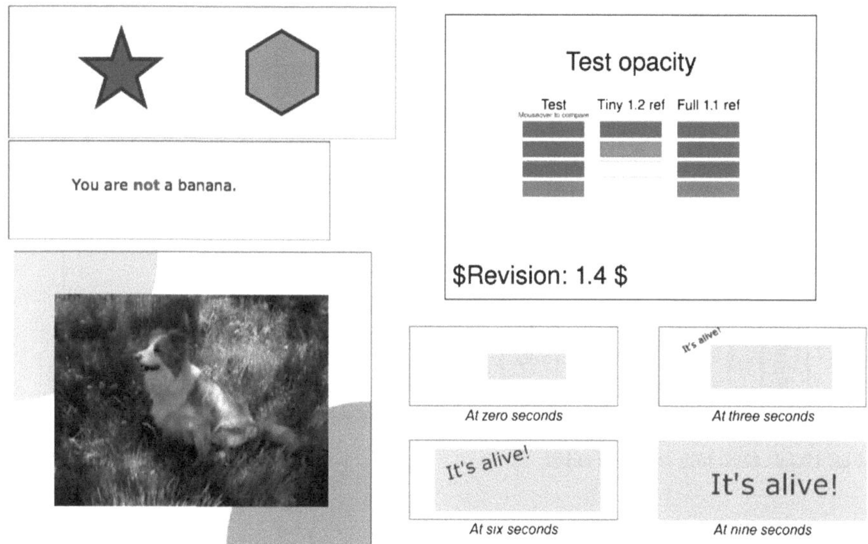

Fig. 10.8 SVG scene description

```
<ev:listener event="accessKey(UP)" handler="#s1"/>
<lsr:conditional begin="s1">
        <lsr:Add ref="g" attributeName="translation" value="0 -5"/>
</lsr:conditional>
<ev:listener event="accessKey(DOWN)" handler="#s2"/>
<lsr:conditional begin="s2">
        <lsr:Add ref="g" attributeName="translation" value="0 5"/>
</lsr:conditional>
```

Fig. 10.9 Example of LASeR "conditional" element

- Selector: to act as a selection element, rendering zero or one of its children, in addition to the semantics of the SVG g element.
- SimpleLayout: to act as a simple layout tool, by spacing its children by a specified amount, thus creating rows or columns of children, in addition to the semantics of the SVG g element.
- SetScroll: a simplified animateScroll.
- StreamSource: to first make available at the scene level information about the state of media chains and secondly to give hints for potential resource optimization.
- Updates: to provide a link to a source of scene updates.

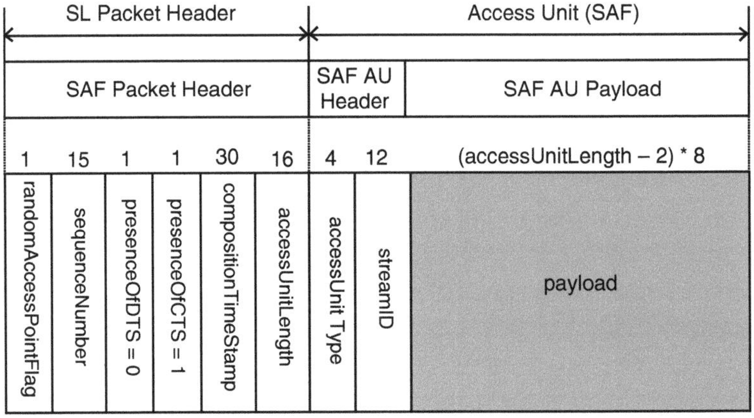

SL Packet Header | Access Unit (SAF)

SAF Packet Header | SAF AU Header | SAF AU Payload

1	15	1	1	30	16	4	12	(accessUnitLength – 2) * 8
randomAccessPointFlag	sequenceNumber	presenceOfDTS = 0	presenceOfCTS = 1	compositionTimeStamp	accessUnitLength	accessUnit Type	streamID	payload

Fig. 10.10 Structure of SAF Packet

10.3.2.3 LASeR Binary Encoding and Transmission

A scene explained in Sect. 3.2.2 is defined in terms of XML. The XML based scene description is suitable for web browser and human readability. However, it has disadvantages in terms of transmission because the XML based description cannot be transmitted by streaming method but only by downloading. Also, the parser for rendering a XML based description requires more processing power than decoding encoded bitstreams. In order to overcome these disadvantages, LASeR provides binary encoding methods of XML based description, which uses fixed-length coding (FLC) for point data, and Exp-Golomb coding to point sequence data.

In order to distribute the encoded scene description data for streaming services, LASeR defines a transmission specification called Simple Aggregation Format (SAF), that is designed for multiplexing multiple encoded scene description bitstreams and component media data into one SAF stream, and transmitting it with relatively smaller header information especially at a lower bitrate suitable for mobile communication channels. As shown in Fig. 10.10, a SAF stream is composed of SAF Access Units (AU), which have the different types shown in Table 10.2.

Since SAF is designed for low bitrate environments such as mobile communication channels, it reduces the header size for multiplexing, and also provides cache functionality, which can store the information frequently accessed or referenced in a memory and thus, minimizes the transmission delay time when a scene is updated. Additionally, SAF supports a variety of media types, including non-MPEG media data types.

Since SAF is defined independently from a network layer, it can be easily transmitted using various network protocols. For example, Fig. 10.11 shows how to transmit SAF data over the RTP protocol, where it is found that the mapping between RTP protocol and SAF packet is very easy and efficient.

Table 10.2 AccessUnitType values and corresponding data in the payload

Value	Type of access unit payload	Data in payload
0x00	Reserved	–
0x01	TransientStreamHeader	A SimpleDecoderConfigDescriptor
0x02	NonTransientStreamHeader	A SimpleDecoderConfigDescriptor
0x03	EndofStream	(No data)
0x04	AccessUnit	An access unit
0x05	EndOfSAFSession	(No data)
0x06	CacheUnit	A cache object
0x07	RemoteStreamHeader	An url and a SimpleDecoderConfigDescriptor
0x08	GroupDescriptor	–
0x09	FirstFragmentUnit	The first Fragment of an Access Unit
0x0A	FragmentUnit	A Fragment of an Access Unit (not the first fragment)
0x0B	SAFConfiguration	A safConfiguration object
0x0C	StopCache	–
0x0D ~ 0x0F	Reserved	–

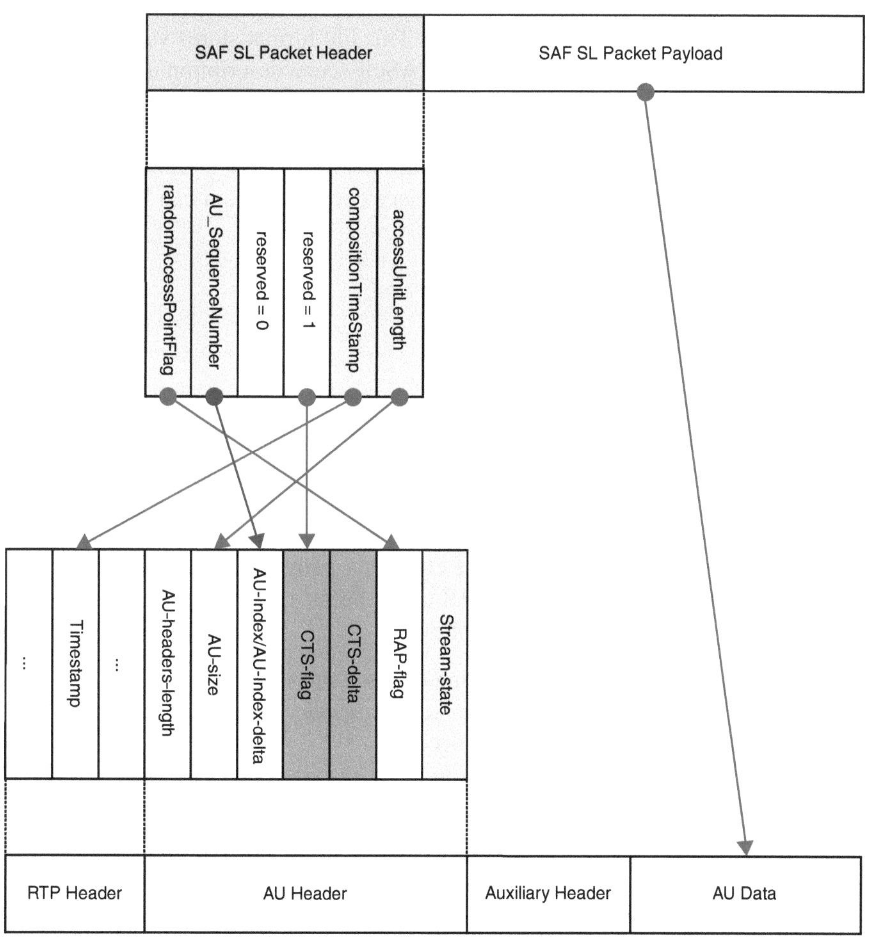

Fig. 10.11 Mapping SAF packet to RTP payload format

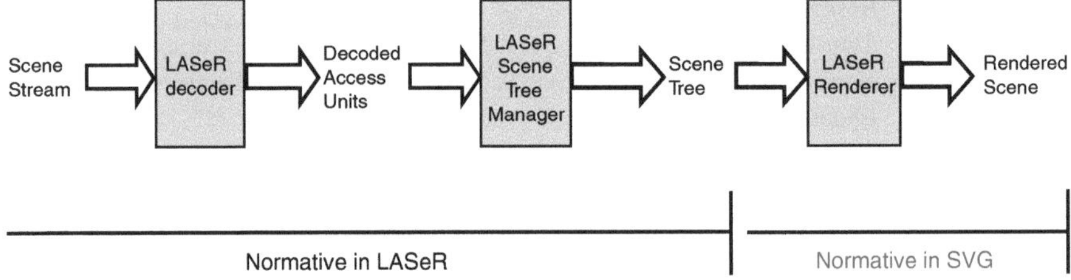

Fig. 10.12 LASeR systems decoder model

10.3.2.4 LASeR Content Storage

LASeR content can be stored according to the MP4 file format standard which is based on ISO base media file format [12]. This file format stores various component media data in individual tracks and LASeR scene description data in a separate scene track in a file. The media type of the track containing a LASeR scene description data is defined in terms of scene description handler_type such as "sdsm". Also, the LASeR header information is stored in a ConfigurationBox under a track header, and thus, LASeR binary encoded stream can be stored in a file format.

10.3.2.5 LASeR Systems Model

Figure 10.12 shows the LASeR systems decoder model, which defines how to decode and render a LASeR content when it is received. The LASeR Systems decoder model provides an abstract view of the behavior of the terminal complying with LASeR contents. It may be used by the sender to predict how the receiving terminal will behave in terms of buffer management and synchronization when decoding data received in the form of elementary streams. The LASeR systems decoder model includes a timing model and a buffer model. The specification of LASeR systems decoder model is as follows:

- The conceptual interface for accessing data streams (Delivery Layer),
- Decoding buffers for coded data for each elementary stream,
- The behavior of elementary stream decoders,
- Composition memory for decoded data from each decoder, and
- The output behavior of composition memory towards the compositor.

As shown in Fig. 10.12, a LASeR stream transmitted by a sender is decoded by a LASeR decoder into an AU, which is a component media data or a LASeR command. These AUs are used for LASeR scene tree structure, which is equivalent

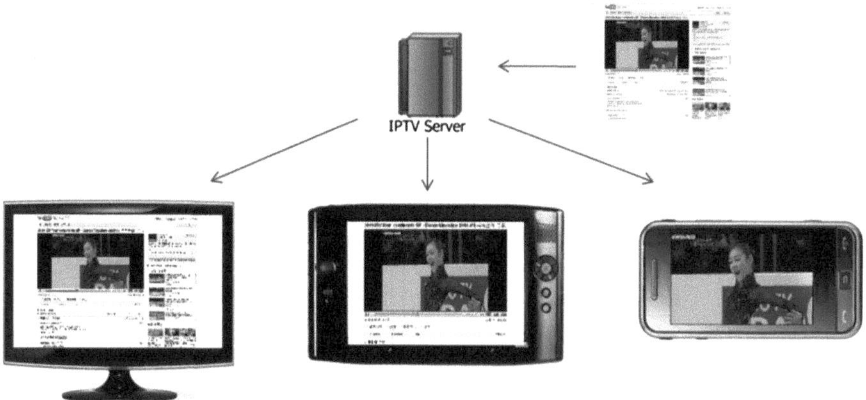

Fig. 10.13 Example of scene adaptation

to Document Object Model (DOM) tree in SVG. This tree structure provides a hierarchy of component media data in terms of spatial and temporal relationship.

10.3.3 Scene Adaptation

As multimedia devices are more widely used, the service scenario of one-source distributing content to different consumption environments becomes real. Especially, the terminals consuming multimedia have differing parameters such as LCD size, CPU speed, memory size and transmission bandwidth. In order to provide customized multimedia services under this environment, LASeR Amendment 2: Scene adaptation was developed and finally approved as an international standard in 2009.

As described above, the purpose of Scene Adaptation is to provide the functionality of adapting a scene description to a specific user environment. For example, when a LASeR content composed of multiple media components is transmitted to various terminals shown in Fig. 10.13, the same content can be differently displayed depending on device conditions. E.g. a larger screen device shows a full content as it is, and a smaller screen presents the same content only partially. That is, a content can be differently consumed depending on individual multimedia devices by using an adaptation of a scene description.

The LASeR Scene Adaptation defines Display Size adaptation, Memory adaptation, Resolution adaptation and Font Resolution adaptation for this service realization. Also, it presents the parsingSwitch element for signaling alternatives in a scene tree with respect to memory requirements, and AdaptiveUpdateGroup command for wrapping LASeR updates with an indication of the associated required adaptation criteria. It can be used for adaptation purposes as shown in Fig. 10.14.

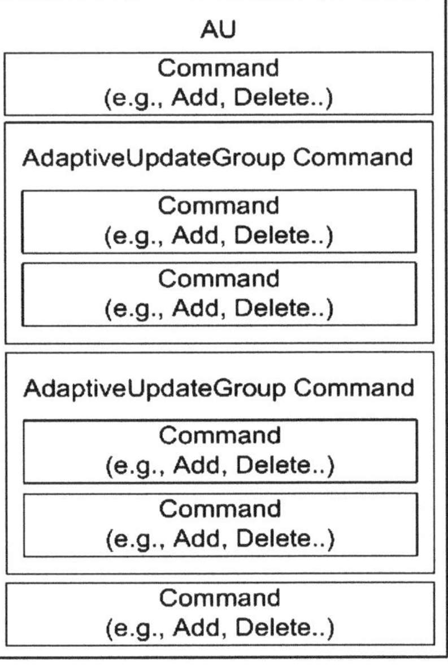

Fig. 10.14 LASeR updates in one LASeR Access Unit (AU)

10.3.4 Presentation and Modification of Structured Information (PMSI)

Presentation and Modification of Structured Information (PMSI) is a recent LASeR activity. PMSI aims at efficiently representing and updating a content scene by referencing partially structured information. In general, richmedia is updated with a scene description including multiple commands. Thus, PMSI considers a richmedia in terms of two aspects: Presentation Information (PI) and Structured Information (SI). PI provides a scene description such as spatial and temporal relationship among component media, and SI is data to be presented as richmedia. In other words, PI and SI have the role of "how" and "what," respectively.

SI is structured information such as metadata in terms of text, for example, MPEG-21 Digital Item Declaration (DID), MPEG-7 metadata and TV-Anytime metadata. Since SI does not have presentation functionality, the same SI can be differently rendered and displayed depending on a scene description. Since richmedia is not designed for transmission of large-size metadata, it is not efficient to transmit richmedia with heavy SI. Therefore, LASeR PMSI separately defines data to be displayed and how to display it, that is, SI for metadata and PI for its presentation, and finally specifies referencing mechanism between those two information types. For example, Fig. 10.15 shows an EPG presented using PMSI, where the way the EPG page looks like is specified by PI, and the contents of the EPG such as TV program schedule is specified by SI. When only part of content in the EPG, such as weather information has changed, the only relevant SI is updated and presented by

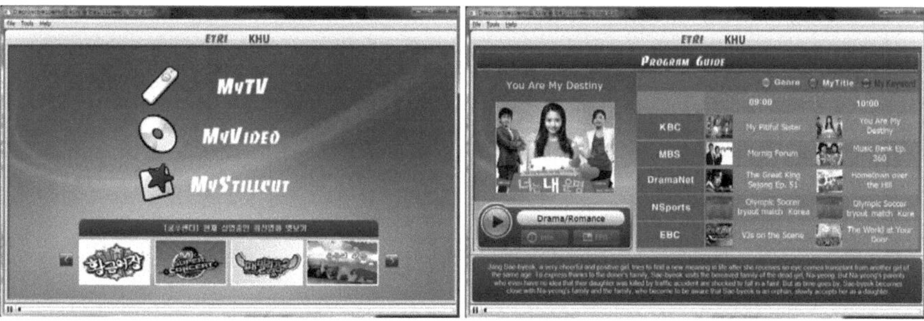

Fig. 10.15 EPG example of PMSI

PI with the referencing mechanism, where it is not needed to update and present all the data in the EPG.

The PMSI specifies the elements for the referencing mechanism as follows:

- Tref: to get SI data itself by using xlink:href
- ExternalReference: to identify the scope of a scene to be updated regularly with the latest version of Structured Information at the specified interval of time such as updateInterval
- ExternalUpdate: to support a modification of Structured Information or elements|attributes thereof. It supports three types of modification: "replace," "insert," and "delete". A part of Structured Information can be replaced with a new element or an attribute value, a new element or an attribute value is inserted into Structured Information, or a part of Structured Information can be deleted according to the provided ***type*** attribute.
- RemoteStreamHeader: to contain ID for a referencing element such as url information of image, video and audio

References

1. ISO/IEC 14496–11, Information technology – Coding of audio-visual objects – Part 11: Scene description and application engine
2. ISO/IEC 14772, Information technology – Computer graphics and image processing – The Virtual Reality Modeling Language (VRML)
3. Khronos Group, Collada Specification, http://www.khronos.org/collada/
4. ISO/IEC 19775, Information technology – Computer graphics and image processing – Extensible 3D (X3D)
5. ISO/IEC 14496–5, Information technology – Coding of audio-visual objects – Part 5: Reference software
6. GPAC Project on Advanced Content, http://gpac.sourceforge.net
7. ISO/IEC 16262, Information technology – ECMAScript language
8. ISO/IEC 14496–1, Information technology – Coding of audio-visual objects – Part 1: Systems
9. ETSI TS 102 427 and ETSI TS 102 428, Terrestrial Digital Multimedia Broadcasting (T-DMB)

10. ISO/IEC 14496–20, Information technology – Coding of audio-visual objects – Part 20: Lightweight Application Scene Representation (LASeR) and Simple Aggregation Format (SAF)
11. Scalable Vector Graphics (SVG)
12. ISO/IEC 14496–12, Information technology – Coding of audio-visual objects – Part 12: ISO base media file format

Chapter 11
MPEG 3D Graphics Representation

Francisco Morán Burgos and Marius Preda

The Tri-Dimensional Graphics (3DG) information coding tools of MPEG-4 are mostly contained in MPEG-4 Part 16, "Animation Framework eXtension (AFX)", and focus on three important requirements for 3DG applications: compression, streamability and scalability. There are tools for the efficient representation and coding of both individual 3D objects and whole interactive scenes composed by several objects. Usually, the shape, appearance and animation of a 3D object are treated separately, so we devote different sections to each of those subjects, and we also devote another section to scene graph coding, in which we describe how to integrate MPEG-4's 3DG compression tools with other (non MPEG-4-compliant) XML-based scene graph definitions. A final section on application examples gives hints on the flexibility provided by the 3DG toolset of MPEG-4.

11.1 Introduction

This chapter focuses on the Tri-Dimensional Graphics (3DG) information coding tools contained in MPEG-4, which was conceived by MPEG as the first set of specifications for the compression of truly multimedia content. Unlike MPEG[-1] and MPEG-2, which only dealt with the coding of natural video and audio (and some related systems information), MPEG-4 was aimed at compressing interactive scenes (so the user stopped being regarded as a mere passive viewer/listener) with different kinds of multimedia objects, including synthetic 3DG ones. It is worth noting that

F. Morán Burgos (✉)
Grupo de Tratamiento de Imágenes, E.T.S. Ing. Telecomunicación,
Universidad Politécnica de Madrid, 28040 Madrid, Spain
e-mail: fmb@gti.ssr.upm.es

L. Chiariglione (ed.), *The MPEG Representation of Digital Media*,
DOI 10.1007/978-1-4419-6184-6_11,

another MPEG standard, namely MPEG-7, contains as well some 3DG-related tools, but they are aimed at describing the 3DG content, not at efficiently coding it, and they are not reviewed below.

11.1.1 Necessity and Basics of MPEG Standards Related to 3DG

Since 1963, when Sutherland created the first computer program for drawing [1], the field of synthetic graphics has followed, as many others involving computers, more or less the same exponential growth foreseen by Moore in 1965 for the semi-conductor industry. In the early years, the progress was pushed up by scientific interest, the real boom starting when the need for special effects and 3DG content came from the film industry, and later from the videogames one. Of those two industries, the former leads the development of technologies for the production of high quality images, and the latter, which has already outgrown the former in terms of market share, that of efficient techniques for real-time processing and rendering.

Even from the beginning, the strategy for developing 3DG solutions was based on three requirements: faster, cheaper, better (image quality). Traditionally, the 3DG chain was focused on the two ends: production and consumption, each side having its own strong requirements. Additionally, in many applications such as games, dealing with complex 3D environments, there is an iterative loop involving content creators and programmers or beta-testers, making sure that the content fits a specific end-user terminal configuration. The complexity of creating 3DG content leads to the coexistence of several authoring tools, each specialized in one or more particular tasks. Therefore, it is possible to (have to) use several tools to create one single 3DG asset, and interoperability is an issue, so either a unique representation standard is used by all tools or data format converters must be provided in each of them, and for all others in the ideal case.

In the last decade, several efforts have been made to develop a unique data format for interchanging assets between authoring tools. In the category of open standards, X3D [2] (the evolution of VRML97 [3]) and COLLADA [4] are the best known, the latter probably being the most adopted by current tools. All these standards specify textual formats providing formalisms both for graphics primitives, to represent the characteristics of isolated 3D objects (their shape, appearance, animation, etc.) and for the scene graph, to describe the space-time relations between different objects, and between them and the user. While COLLADA concentrates only on representing (static) 3D objects or scenes, X3D pushes the standardization further by addressing as well user interaction thanks to an event model in which scripts, possibly external to the main file containing the 3D scene, may be used to control the behavior of its objects.

On the consumption side, data should be as "flat" (i.e., close to the hardware) as possible to enable easy and rapid processing by the graphics cards. A lot of effort has been spent for optimizing the rendering step, one of the bottlenecks in the 3D synthetic content pipeline. Better, faster algorithms are continuously proposed and some of them have been directly implemented in the silicon, allowing rendering performances that some years ago nobody could have even dreamed of. To normalize somehow the access to different graphics cards, low level 3DG Application

Programming Interface (APIs) like OpenGL and Microsoft Direct3D were created and keep evolving. Such APIs hide the different capabilities of the many and wildly diverse graphics cards in the market (with and without specific hardware acceleration for 3D visualization, with more or less texture memory, etc.) and offer the programmer a single set of rendering functions.

Recent developments of collaborative sharing platforms are efficient vehicles for pushing media over the Internet. By potentially transforming every end user in a content producer, Web2.0 platforms lead to the proliferation of media content, therefore introducing an even greater need of compactness in data representation. Even in the case of games, which have traditionally been distributed in DVDs containing all assets, openness to new distribution channels is a must, either for entire game download before playing, or for the streaming of assets during the match in the case of on-line games. Neither the formats used in the content creation phase, nor the ones used for rendering are appropriate for transmission, since they usually satisfy none of the three requirements that are most important in this context: (1) *compression*, to reduce the size of the transmitted data; (2) *streamability*, to enable players to start using the content before having downloaded it entirely; (3) *scalability*, to ease real-time adaptation of the content to a specific platform and network.

MPEG-4 was also built on top of VRML97 and contained, already in its first two versions published by ISO in 1999 and 2000, tools for the compression and streaming of 3DG assets, enabling to describe compactly the geometry and appearance of generic, but static objects, and also the animation of human-like characters. Since then, MPEG has kept working on improving its 3DG compression toolset and published already three Editions of MPEG-4 Part 16, "Animation Framework eXtension (AFX)" [5], which addresses the requirements above within a unified and generic framework.

11.1.2 Chapter Structure and Location of 3DG Tools in MPEG-4

As any good book on computer graphics [6] explains, to model and later compress a synthetic 3DG scene containing several objects, and their evolution over time, and how the user may interact with them, two separate problems have to be solved: the representation of *each individual 3D object* and that of *the whole, interactive 3D scene*. All standards for synthetic 3D (or even 2D) scene modeling and compression segregate those two problems, and MPEG-4 is no exception. Section 11.2 introduces several basic concepts necessary to understand the 3DG compression tools of MPEG-4, which have been published in several parts of the standard, and are explained in more detail in Sect. 11.3–11.6:

- In turn, to model and compress a single 3D object, even if it is *static* (it does not move and is not deformed over time), two very different kinds of information must be specified to describe its *shape* and *appearance*, and this is done again separately in most cases. Section 11.3 reviews the most important tools normalized by MPEG-4 for shape compression, whereas Sect. 11.4 is devoted to appearance compression tools. Furthermore, if the object is *dynamic*, its *animation* over time must be specified as well: see Sect. 11.5. Most compression tools for individual

3DG objects belong in MPEG-4 Part 16, "Animation Framework eXtension (AFX)" [5]. However, MPEG-4 Part 2, "Visual" [7], contains some old tools such as 3D Mesh Coding (3DMC), for the compression of generic, but static meshes (see Sect. 11.3.1.1), and the Face and Body Animation (FBA) toolset, for the compression of dynamic humanoids (see Sect. 11.5.1).
- Representing and coding the interactive scene means specifying the interrelations, over space and time, of objects between themselves and with the user. The essential concept here is that of *scene graph*: see Sect. 11.6. Tools related to object and scene graph representation are mostly in Part 11, "Scene Description and Application Engine" [8], and is the current home of MPEG-4's BInary Format for Scene (BIFS) toolset, which was originally published in Part 1, "Systems" (see Chap. 10), and did not only inherit from VRML97 the scene graph concept, but also many scene nodes (e.g., IFS), and generic animation mechanisms (e.g., position and orientation interpolators). Part 25, "3D Graphics Compression Model" [9] offers an architectural model able to accommodate a third party XML-based description of the scene graph with potential binarization tools and with MPEG-4 3DG compression tools.
- Finally, like all others in MPEG-4 (and other MPEG standards), 3DG-related tools have reference software [10] and conformance bitstreams [11] associated to them.

11.2 3DG Compression Basics

11.2.1 Static Object Coding

Very few 3D object modeling and coding techniques pay attention to the object inner volume, especially if what is most important is the visual result once the object is subsequently decoded and rendered. Since only the object surface shape and appearance are taken into account by most rendering engines, "truly 3D" models describing solids are much less frequent than "merely 2,5D" ones describing surfaces (immersed in 3D space, but with just two degrees of freedom). We will therefore focus on 3D *surface* modeling.[1]

As for how the *shape* and *appearance* of a surface are described, the traditional approach, which deals with them separately, remains the most common by far. There do exist hybrid Image-Based [Modeling and] Rendering (IB[M]R) techniques to represent simultaneously those two kinds of information, and they do have certain advantages over the traditional approach, but they are not as mature as the latter, which is still the only one for which current Graphics Processing Units (GPUs) provide obvious hardware acceleration support. We will therefore focus on traditional

[1] AFX contains tools for the efficient coding of solids, and of surfaces approximated by Bézier's or NURBS patches, and for the joint description of the shape and appearance of a 3D object, such as the ones contained in the Depth Image Based Representation (DIBR) and DIBR version 2 toolsets, but we will not address them Fig. 11.1.

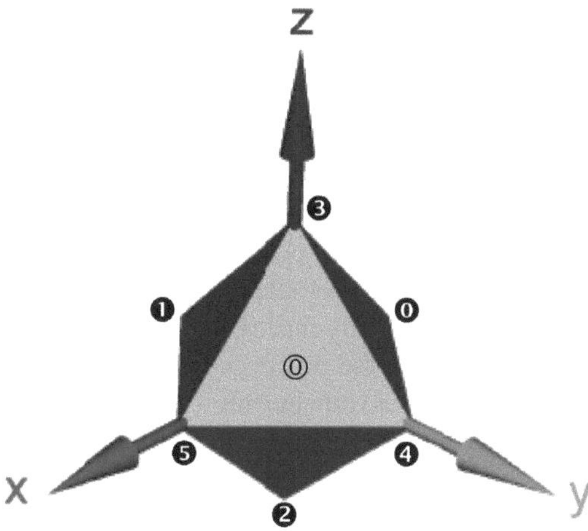

Fig. 11.1 IFS representation of an octahedron

surface representation techniques, in which the shape is explicitly described first, and the appearance linked to it later.

This leads us to another taxonomy concerning the shape approximation degree. Higher-order, polynomial or rational patches, such as Bézier's or Non-Uniform Rational B-Spline (NURBS) ones, are certainly very used in Computer-Aided Design (CAD) [12], and greatly simplify the edition of synthetic 3D models. But in the "wider 3DG world", especially if the videogame industry is included in it as it should be today, the most common surface approximation is still the piecewise linear one: indeed, unsophisticated as it may sound, tiling the (likely smooth) desired surface with a mesh of (flat, by definition) polygons is the preferred option. The main reason for this is again that the current rendering engines and GPUs provide specific support for polygon meshes. We will therefore focus on the representation and compression of *triangle meshes*, since triangles are the simplest polygons, and all other can be triangulated.

A 3D triangle mesh is almost invariably represented with two tables, as illustrated by Fig. 11.1: one for *geometry*, in which the three coordinates (x, y, z) of each vertex are given; and another for *topology* (or *connectivity*), listing the indices (i, j, k) in the previous table of the three vertices forming each triangle. This pair of tables is commonly known as an Indexed Face Set (IFS), which is precisely the name (except for the spaces) of a VRML97 node. Given the clear distinction between the geometry information and the topology one, most static 3D triangle mesh compression techniques [13] treat them separately.

As already stated above, in traditional surface representation techniques, the shape is described first, and then an appearance is linked to it. For a triangle mesh, this is done by assigning *attributes* to its elements (triangles, vertices, etc.: see next paragraph). Examples of such attributes are a color, hence a 3D vector in the Red, Green, Blue (RGB) color space; or a local normal vector to the surface, hence again a 3D vector, although in the usual (x, y, z) space; or, most typically, a pair of texture coordinates, hence a 2D vector in the (u, v) texture space, pointing to a pixel (hence an RGB color) in some texture/image that is to be wrapped around the shape at rendering time through indirections.

Attribute mapping is specified at modeling time, and may follow a per-triangle approach: for instance, a single color or normal vector may be assigned to an entire triangle and then it will be used, at rendering time, to fill all the pixels covered by that triangle after performing the necessary lighting calculations, if any. However, the resulting "faceted" rendering style is not photorealistic, so attributes are much more often mapped onto the mesh on a per-vertex basis, or even on a per-corner one, a corner being a vertex considered within a particular triangle. In particular, texture coordinates are always mapped on a per-vertex/corner fashion. At rendering time, and more specifically during the rasterizing process, some kind of interpolation is performed to fill the pixels covered by each triangle after performing the lighting calculations at its vertices only.

Coding the appearance of a 3D triangle mesh (see Sect. 11.4) involves of course image compression techniques if textures are used, as is often the case, to achieve a more photo-realistic result. But it involves as well, even if textures are not used, attribute coding techniques, which are not too different from the geometry coding ones when attributes are mapped on a per-vertex basis: a vertex has then, on top of the obvious (x, y, z) 3D vector giving its position, other information attached to it, such as an (r, g, b) or (n_x, n_y, n_z) 3D vector, or a (u, v) 2D one.

11.2.2 Dynamic Object Coding

Some of the mesh features mentioned above may be continuously updated, therefore transforming the static object into a *dynamic* one. Most commonly, two kinds of features are animated over time, and both concern the geometry data: either the position and/or orientation of the whole set of vertices is changed globally, which means that the object moves and/or rotates inside the scene, or the locations of some vertices are altered relative to that of others, which yields a shape deformation. On the other hand, mesh connectivity remains almost always consistent over time, and so do textures and attributes.

Mesh deformation coding is addressed in two ways: one can either code the new vertex positions themselves as a function of time, or use some deformation controller to influence the mesh geometry, and code the controller parameters as a function of time. In both cases, classic signal processing techniques such as space-frequency transforms or predictive and entropy coding can be used to reduce the data size. Usually, the animation data is highly redundant, so compressing it may easily reduce the size by two orders of magnitude.

11.2.3 Scene Coding

The *scene graph* is the general data structure commonly used by vector-based graphics editing applications and modern computer games. This structure arranges the logical and spatial representation of a multimedia scene and corresponds to a collection of nodes organized as a *hierarchical graph* or tree, whose nodes may have many children but only one single parent. An operation applied to a parent node automatically propagates its effect to all of its children. In many scene graph implementations, associating a geometrical transformation matrix at each level and concatenating such matrices together is an efficient and natural way to perform animation. A common feature is the ability to group related objects into a compound object which can then be transformed, selected, etc., as easily as a single object.

Historically, the scene graph concepts are inherited from techniques enabling to optimize 3DG rendering. In order not to process invisible objects (i.e., objects outside the window chosen by the user), a logic organization of relations between them is created. Additionally, common properties such as textures or lighting conditions may be used to group together different objects; consequently, loading and unloading operations per object are now perform per group of objects.

Because of the heterogeneity of the scene graph, encoding it will not allow high compression rates (10:1 is common), only small redundancy being exploitable, especially thanks to predictive and entropy coding. Quantization is difficult to use without knowing the semantics of the data to be encoded and the requirements with respect to the reconstruction accuracy.

11.3 Shape Compression Tools

As in the cases of image or video coding, the concepts of resolution scalability and of *single*-vs. *multi-rate coding* are important for 3D mesh coding. In some scenarios, a single-rate coding of a 3D mesh may be enough, but its progressive compression may be desirable, especially if it is a complex mesh to be transmitted over a network with a restricted bandwidth, or to terminals with limited processing power. In such cases, it is useful to represent and code the original, fine mesh as a sequence (or a hierarchy) of refinements applied to a simple, coarse mesh. During decoding, connectivity and geometry are reconstructed incrementally from the bitstream until the original mesh is rendered at its full resolution or the transmission is cancelled by the user. Progressive compression thus allows transmission and rendering of different Levels Of Detail (LODs).

11.3.1 Single-Rate Coding

To establish the compression efficiency of the tools reviewed in this section, let us start by giving an estimate of a *triangle mesh raw size* (before compression), S_{raw}. Following the IFS paradigm described in Sect. 11.2.1, and supposing a 32-bit float is used for each coordinate in the geometry table and a 32-bit integer for each index in the topology one, a first estimate of S_{raw} for a mesh with V vertices and T triangles is $S_{raw,\, naive} = 3 \cdot 32\,(V + T)$ bit. Since typical triangle meshes, and especially large ones, have a vast majority of regular vertices (i.e., vertices connected to another six), their number of triangles is approximately twice that of vertices, so S_{raw} can be expressed exclusively in terms of V. In fact, it is common practice in the mesh coding literature to express bitstream sizes in *bits per vertex* (*bpv*), in the same way it is usual to measure the size of a compressed image in bits per pixel (bpp). With the assumptions above, $S_{raw, naive} = 288$ bpv. However, a more reasonable estimate can be obtained by arguing that:

- *Geometry* data may be coded in a *lossy* way, so 32 bits per coordinate are far too many: even with uniform quantization, 16 bits are enough to resolve 15 μm details in a model of a human body, or 1 mm ones in a model of a big building.
- *Topology* information, however, must be coded in a *lossless* way most usually (and definitely so if the original IFS must be reconstructed exactly, except perhaps for a certain tolerance in the position of its vertices), but 32 bits are equally too many for the vertex indices: $3 \lceil \log_2(V) \rceil T$ bits should suffice, even without applying any sophisticated topology coding scheme.

Therefore $S_{raw} = 3\,(16V + \lceil \log_2(V) \rceil T)$ bit $\cong 3\,(16 + 2\lceil \log_2(V) \rceil)$ bpv. For a relatively large mesh with, say, 50 K vertices (more precisely, up to $2^{16} = 65{,}536$), and hence around 100 K triangles ($2^{17} = 131{,}072$), $S_{raw, large\ mesh} = 144$ bpv. One third of this budget is devoted to geometry and two thirds to topology.

11.3.1.1 The Past: 3D Mesh Coding [Extension] (3DMC[e])

The problem of compactly encoding the connectivity information of a 3D polygon mesh has been studied extensively, starting with the theory of planar polygon graphs. Tutte established in 1962 that, in the case of arbitrary triangle graphs, their encoding consumes at least $\log_2(256/27) \cong 3.245$ bpv, but this limit does not apply to large meshes with a majority of regular vertices such as the ones defined above. In fact, Turán concluded in 1984 that a planar graph topology can be encoded with a constant number of bpv using two spanning trees: a vertex spanning tree and a triangle spanning tree. This led Taubin and Rossignac to invent in 1998 their *Topological Surgery* (*TS*) scheme [14] for 3D mesh connectivity coding.

The idea is to cut a given 3D mesh along a selected set of edges to make a planar polygon, as illustrated by Fig. 11.2, which shows how the same mesh of 4·20 triangles (an icosahedron whose triangles have been subdivided into four each) can lead to two different triangle trees depending on the set of cut edges (i.e., the vertex tree). The mesh connectivity is then represented by both trees, yielding 1 bpv for very regular meshes and 4 bpv on average otherwise.

TS is the basis for the old 3D Mesh Coding (3DMC) toolset of MPEG-4 Part 2, and also for the 3DMC eXtension (3DMCe) [15] tool of AFX, which adds support for efficient texture coordinate compression, and for mesh animation/editing.

As for geometry information, to achieve higher compression rates than the 3·16 bpv mentioned above, both 3DMC and 3DMCe feed the vertex coordinates to a *quantization* step, whose resulting values are then compressed by entropy coding after some

Fig. 11.2 Two examples of topological surgery in which the same mesh is cut through two different edge sets (the *vertex trees*) to form two simply connected polygons, whose dual graphs are the two corresponding triangle trees (taken from Ref. [14])

prediction (relying on some smoothness assumptions) is applied. Both steps contribute to the compactness of the final result, but quantization is intrinsically and irreversibly lossy, whereas prediction is a perfectly reversible and lossless transformation of the signal to make it fit for a more efficient subsequent entropy coding. Vertex coordinates are typically uniformly quantized with 8-14 bits each and prediction is usually linear: at most three vertices adjacent to the one being decoded, and already decoded themselves, according to the vertex ordering imposed by TS, are used for the prediction known as the "parallelogram rule". This yields bitrates of some 13-18 bpv for 9-12 bits per coordinate, or 13 bpv at 8-bit quantization resolution.

11.3.1.2 The Present: Scalable Complexity 3D Mesh Compression (SC3DMC)

The Scalable Complexity 3D Mesh Compression (SC3DMC) toolset of AFX specifies a way to fine tune the trade-off between compression efficiency (bpv) and computational resources (CPU and memory) needed in both encoder and decoder by choosing among three parameterized 3D mesh coding techniques: Quantization-Based Compact Representation (QBCR), Shared Vertex Analysis (SVA) and Triangle FAN (TFAN).

The main idea behind this scalable complexity 3D mesh compression scheme is that, in some application scenarios, especially the ones involving mobile devices with reasonable network connections, the minimization of bitstream size may not be as important as that of computational resources. Current implementations in the reference software of this AFX Amendment yield encoding performances of 17-70 bpv with associated decoding speeds of 1-3 millions of vertices per second on an ordinary PC.

11.3.2 Multi-resolution Coding

The concept of *Progressive Mesh* (*PM*) [16] allows to code a mesh with a total of around 35 bpv, including both topology and geometry information. A PM is a base mesh plus a sequence of vertex split records, each specifying which vertex and pair of edges incident to it must be split, and the local geometry changes. From such a representation, it is straightforward to extract a LOD of the mesh with any desired number of triangles by simply choosing the adequate prefix of the vertex split sequence, which is streamed after the base mesh has been transmitted.

11.3.2.1 The Past: Progressive Forest Split (PFS), Wavelet Subdivision Surface (WSS), MeshGrid and FootPrint

The Progressive Forest Split (PFS) technique [17] is based on the PM and TS ideas, and was included in the 3DMC toolset of MPEG-4 Part 2. PFS is able to reduce the bitrates of PMs at the expense of reduced granularity: two successive LODs of a PFS set differ by a group of vertex splits, instead of only one. Logically enough, the highest compression ratios are achieved by minimizing the number of LODs but, typically, it is possible to remain slightly below 30 bpv for medium size meshes coded with several LODs.

Traditional progressive coders aim to eventually recover the exact original connectivity. For small meshes with carefully laid out connectivity and sample locations, this is very appropriate. The situation is different for highly detailed, densely sampled meshes coming from 3D scanning: since distortion is measured as geometric distance, the sample locations and connectivity can be treated as additional degrees of freedom to improve the rate-distortion performance. As long as the final result has a geometric error[2] similar to that of the original approximation, the actual sample locations and connectivity do not matter.

Fig. 11.3 shows how an arbitrary connectivity mesh (top left), resulting from an equally arbitrary sampling of a smooth surface, can be remeshed to have "semi-regular connectivity" (top right), so that it may be fed to a truly *hierarchical* (as opposed to merely progressive) 3D mesh coder based on Wavelet Subdivision Surfaces (WSSs) [18, 19]. The connectivity of the semi-regular mesh can be efficiently encoded, as it only depends on that of its underlying base mesh and number of recursive subdivisions. The geometry information is that of the base mesh, plus a

[2] 3D mesh reconstruction errors are typically measured with both the L^2 (i.e., RMS) and $L^{\square}$ (i.e., Hausdorff's or maximum) distances, and expressed in sub-units (10^{-4} is common) of the original mesh bounding box diagonal.

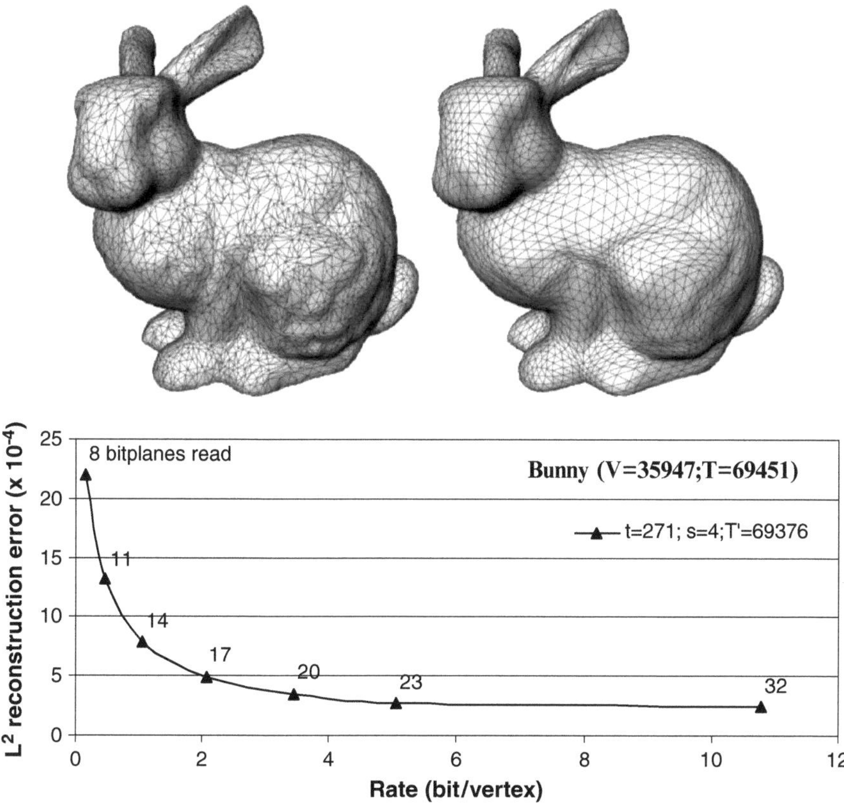

Fig. 11.3 WSS-based coding of the 3D mesh of the "Stanford bunny". *Top*: remeshing of the original, arbitrary connectivity mesh of $T = 69{,}451$ triangles into a semi-regular connectivity one of $T' = t \cdot 4^s = 69{,}376$ triangles ($t = 271$ is the number of triangles of the base mesh, recursively subdivided $s = 4$ times into 4 each). *Bottom* (taken from Ref. [19]): rate vs. distortion curve

hierarchical set of 3D details representing the differences (prediction errors) between successive LODs. The distribution of the coefficients resulting from a space-frequency wavelet transform of these 3D details is centered around zero, and their magnitude decays at finer levels with a rate related to the smoothness of the original surface. This behavior of the wavelet coefficients magnitude is the key to hierarchical coding and justifies the choice of a zerotree coder. The coding efficiency results of the WSS tool in AFX, in terms of reconstruction error for a given bitrate, are over four times better than those of PFS, as can be inferred from the rate-distortion curve of Fig. 11.3 (bottom).

Another hierarchical 3D mesh compression tool of AFX is MeshGrid [20], a compact representation attaching a description of the connectivity between the vertices on the surface, called "connectivity wireframe", to a regular 3D grid of points, called "reference grid". MeshGrid spends very few bpv for topology information thanks to a 3D extension of Freeman's chain-code and to the particular properties of the connectivity wireframe, which allows the decoder to derive the triangulation unambiguously without having encoded it explicitly à la IFS. As for geometry information, the reference grid is a smooth vector field defined on a regular discrete 3D space, and is also efficiently compressed by using an embedded 3D wavelet-based multi-resolution intra-band coding algorithm, not unlike the one used for 3D details in WSS-based coders.

Finally, AFX's FootPrint toolset is yet another solution for the multi-resolution representation of 3D objects based on footprints, of which the most notable example are buildings in virtual cities [21]. Indeed, the on-line, interactive navigation over a large and densely built urban environment raises difficult problems due to the scene size and complexity, and its hierarchical representation allows to build client–server applications in which only the details of particular Regions Of Interest (ROIs) must be streamed by the server to each client, depending on the specific viewpoint chosen in real time by the corresponding user (note that WSSs and MeshGrid are also suited for view-dependent, ROI refinement). FootPrint efficiently exploits the fact that most automated urban environment modeling techniques are essentially extrusion-based and proposes a procedural representation for roofs and façades.

11.3.2.2 The Future: Multi-resolution 3D Mesh Compression (MR3DMC)

One of MPEG's newest activities is that of Multi-Resolution 3D Mesh Compression (MR3DMC), currently in the Core Experiment (CE) phase. The goal of this CE is to design a progressive coding technique for potentially very large 3D meshes of arbitrary topology, that would provide efficient rate-distortion performances while supporting the following functionalities:

- Lossless connectivity coding: the original connectivity must be retrieved exactly, except for a possible permutation of the vertex/triangle indices.
- Spatial scalability: the mesh resolution (LOD) must be globally adaptable to the available terminal rendering performance and network bandwidth.
- View-dependent scalability: it must also be possible to refine locally the mesh resolution for a specific, possibly view-dependent, ROI.
- Quality scalability: the precision of vertex coordinates (and attributes, if any) must be progressively refined as the bitstream is decoded, and the maximum error permitted, once the entire bitstream is decoded, must be controlled by a set of encoding parameters.
- Support of multiple attributes per vertex: for example, texture coordinates, normals, etc. (see as well Sect. 11.4.2).

11.4 Appearance Compression Tools

As already explained in Sect. 11.2.1, at modeling time, an appearance may be linked to a 3D shape through the process of attribute mapping. In the particular but most common case of texture mapping, this enables graphic artists to simulate the wrapping of each 3D object in one or more textures, i.e., 2D images, which have of course to be stored or transmitted along with the information corresponding to the 3D shape and the attribute/texture mapping. This is why we devote the first section below to image coding standards related to MPEG-4, although the most important part of this section, conceptually, is the following section, which deals with MPEG-4's attribute compression tools.

11.4.1 Image Coding

There are many *image compression standards* to choose from, the most commonly used being probably JPEG [22], which can achieve a very high compression ratio (in the order of 50:1 or more) without significant visual quality loss for most natural images. JPEG uses predictive and entropy coding techniques, but these are lossless and cannot provide savings of orders of magnitude. The main reason behind these high compression ratios is the quantization step that follows the (lossless itself) space-frequency transform used by JPEG, the well known Discrete Cosine Transform (DCT). This quantization step is lossy, and variably so (as chosen by the user), but is cleverly designed take good care of the Direct Current (DC, i.e., low/base frequency) coefficient of the DCT, which captures most of the energy of a natural/smooth image, and to destroy selectively the increasingly Alternating Current (AC, i.e., high frequency) coefficients, which normally "contain" much less energy and, besides, are less important for the human visual system.

JPEG is not natively supported by MPEG-4, which has its own image coding tool, namely Visual Texture Coding (VTC) [7], but may be used as the format for a texture in an MPEG-4 scene as an "external" elementary bitstream. On the other hand, JPEG 2000 [23] is supported natively by MPEG-4, and is based, like VTC, on a wavelet-based space-frequency transform, which makes it inherently suitable for multi-resolution decoding of embedded bitstreams. Fig. 11.4 illustrates an example application in which this property would allow the view-dependent, ROI refinement of the 2D texture mapped onto a 3D object [24].

11.4.2 Attribute Mapping Coding

11.4.2.1 The Past: Indexed Face Set (IFS)

In the IFS that MPEG-4 inherited from VRML97, attribute mapping may be specified on a per-face (i.e., per-triangle, if all facets/polygons are triangles), per-vertex or per-corner fashion, but the choice is a global one for the entire IFS. Attribute coding data benefits from exactly the same compression sources as the geometry data (namely, coordinate quantization and entropy coding).

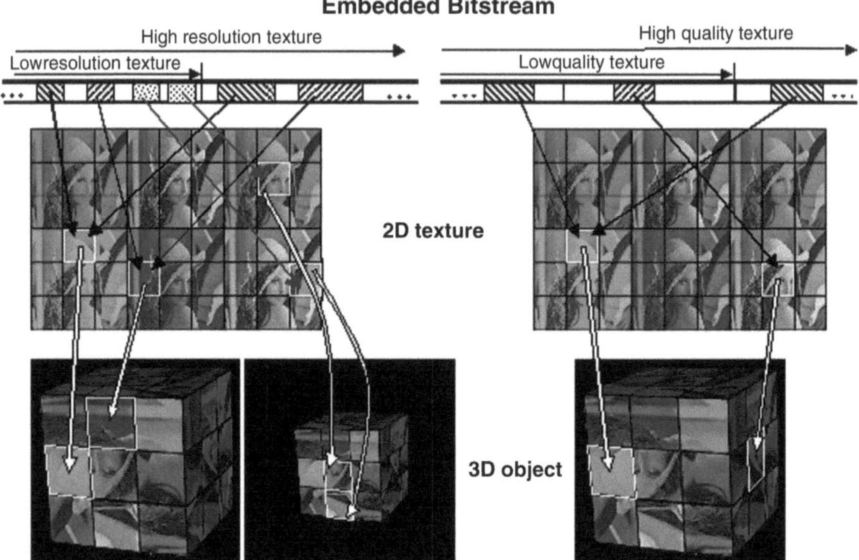

Fig. 11.4 View-dependent texture decoding. Non-contiguous portions of the 2D texture bitstream can be independently extracted and decoded according to the 3D object distance (*left*) and viewing angle (*right*) (taken from Ref. [24])

11.4.2.2 The Future: Indexed Region Set (IRS)

The already finalized and soon to be published Amendment 1 to AFX Edition 4, "Efficient representation of 3D meshes with multiple attributes" [25], introduces the concept of Indexed Region Set (IRS), and the corresponding set of MPEG-4 nodes, to get rid of the limitation of the IFS node inherited from VRML97, which only allows that there be one image to which texture coordinates refer.

Regions in an IRS may be used, for instance, to group sets of faces of (what would have been) an IFS in such a way that each group is assigned a different texture image. Then, region border vertices may be assigned different texture coordinates on a per-corner basis, whereas vertices which are completely interior to a region (i.e., the ones which do not sit on texture borders) need not. This flexibility is not present in the regular IFS node, for which the per-corner specification turns out to be quite expensive.

11.5 Animation Compression Tools

Elaborating models that allow to represent in a compact way the animation data can be fulfilled by exploiting the *redundancy of animation*: (1) *temporal interpolation* is a technique that keeps only the relevant pair of key and key values such as at any moment in time the signal value can be obtained by linear or higher order interpolation;

and (2) *spatial interpolation* refers to clustering vertices and attach to each one a unique transform (e.g., skeleton-based animation). In addition, one can use compression on top of animation to reduce the amount of transmission data. In general, very little research has been conducted in this area. MPEG standardized an approach for compression of generic interpolated data [26], able to represent coordinates and normals interpolation. While generic, this approach does not exploit the spatial redundancy. To do so, MPEG standardized in 2009 a tool called Frame-based Animated Mesh Compression (FAMC) [27] able to cluster vertices based on their motion properties. Concerning avatar animation, one of the most used animation content used in games, MPEG-4 already contained in its first two versions, published by ISO in 1999 and 2000, the Face and Body Animation (FBA) toolset [7, 28, 29] targeted at the compression of virtual humanoids at very low bit-rate, and highly inspired from video coding algorithms. The limitations of FBA in terms of realism made MPEG issue a more powerful and generic toolset, called Bone-Based Animation (BBA), which was originally published in 2004 within the first Edition of AFX, and consists in a multilayer modeling of the avatar and mesh deformation driven by a set of 1D controllers (bones and muscles).

11.5.1 Generic Animation

The expression "generic animation" refers to a representation based on sampling (regularly at each frame or irregularly at key frames) the motion signal. The latter represents data such as positions, orientations or scale factors and can be applied to any object or parts of it from the scene. Typically *rigid object motion*, cameras and lights trajectories may be represented by this technique.

11.5.1.1 The Past: Interpolator Compression (IC)

Key frame animation, and implicitly the interpolation between key frames, is currently the most popular method for computer animation. The interpolator data consist of *key and key value pairs*, where a key is a time stamp and a key value the corresponding value to the key. The data type (vertex coordinates, object orientation or position) determines the dimension of the key value: the key value for vertex coordinates and object position is the set of x, y and z (3D), while the object orientation is represented as an axis and an angle (4D).

The IC structure is illustrated by Fig. 11.5. First, the original interpolator data may be reduced through an analyzer with the role of removing the redundant or less meaningful keys from the original set. A redundant key is defined as a key that can obtained from its neighbors by linear interpolation and a less meaningful key is one for which the distortion between the original signal and the one reconstructed by interpolation is below a given threshold.

Once the significant keys are selected, each component of a pair (key, key value) is processed by a dedicated encoder. The key data, an array with monotonically

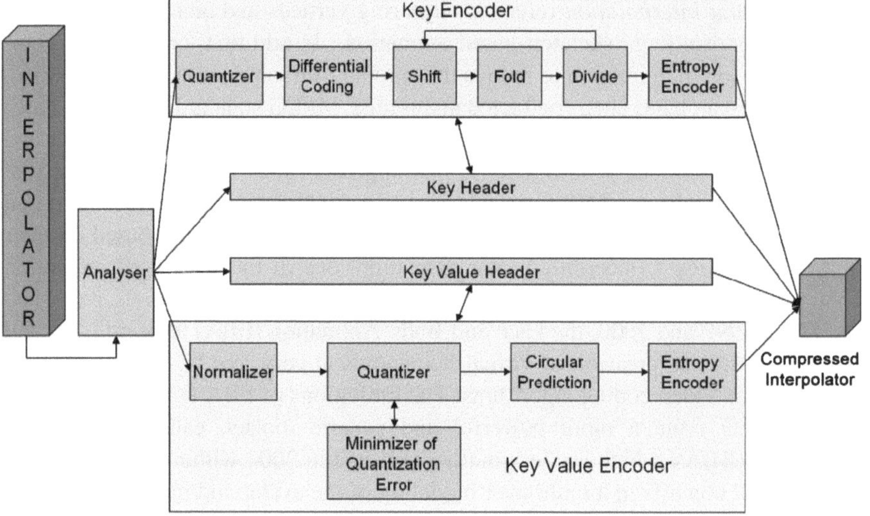

Fig. 11.5 Interpolator encoder

non-decreasing values and usually unbounded, is first quantized and a delta prediction is applied. The result is further processed by a set of shift, fold, and divide operations with the goal of reducing the signal range [26].

Data such as number of keys and quantization parameters are encoded by a header encoder using dictionaries. For the key values, the data is first normalized within a bounding box and then uniformly quantized. The quantized values are predicted from their one or two already transmitted key values and the prediction errors are arithmetically encoded. By using this method for representing interpolators, the compression performances are up to 30:1 with respect to textual representation of the same data.

11.5.1.2 The Present: Frame-Based Animation Mesh Compression (FAMC)

FAMC is a tool to compress an animated mesh by encoding on a time basis the attributes (positions, normal vectors, etc.) of the vertices composing a mesh, independently on how animation is obtained (deformation or rigid motion). The data in a FAMC stream is structured into segments of several frames that can be decoded individually. Within a segment, a temporal prediction model, used for motion compensation, is represented.

Each decoded animation frame updates the geometry and possibly the attributes of the 3DG object that FAMC is referred to. Once the mesh is segmented with respect to the motion of each cluster, three kinds of data are obtained and encoded as shown in Fig. 11.6. First, the mesh partitioning indicates for each vertex the attachment to one or several clusters. Then, for each cluster, an affine transform is encoded by

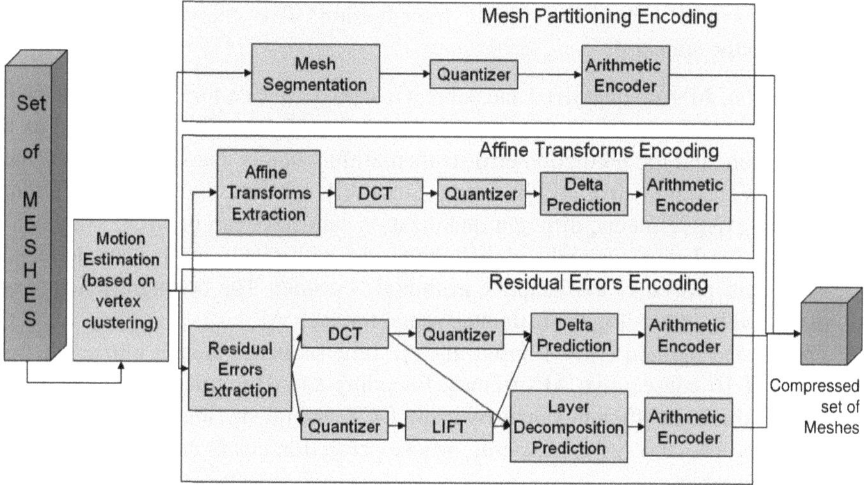

Fig. 11.6 FAMC encoder

following a traditional approach (frequency transform, quantization, prediction of a subset of the spectral coefficients and arithmetic coding). Finally, for each vertex, a residual error is coded. The frequency transform may be the DCT or any from the (possibly lifted) wavelets family (represented as "LIFT" in Fig. 11.6), and the prediction may be the simple delta one, or another based on multi-layer decomposition of the mesh. The standardized implementation [27] of the FAMC encoder achieves compression ratios of up to 45:1 with respect to a textual representation.

11.5.2 Avatar Animation

Virtual characters are the most complex objects in 3D applications and, in order to represent compactly their animation, spatial and temporal redundancy must be exploited. In the first case, due to the biomechanical characteristics, vertices can be clustered and a unique value or geometric transform is assigned to each cluster. In the second, the high correlation of the motion is exploited by frequency domain analysis. Aligned to the hardware capacities of their time, two set of tools were standardized by MPEG: FBA and BBA.

11.5.2.1 The Past: Face and Body Animation (FBA)

The generic nature of the Animation Parameter (AP) description makes it possible avoid transmitting the avatar model during the animation stage. In addition, AP encoding ensures a very low bitrate transmission. Two encoding methods are included in the standard:

1. In the first, APs are quantized and subject to a predictive coding scheme: for each parameter to be coded in frame *n*, its decoded value in frame *n*-1 is used as a prediction, and the prediction error is then arithmetically encoded. This scheme prevents encoding error accumulation. Since APs can be assigned with different accuracy requirements, different quantization step sizes can be used. They consist of a local step size and a global one used for bit-rate control. The quantized values are passed to the adaptive arithmetic encoder. The coding efficiency is increased by providing the estimated range for each AP.
2. The second method is DCT-based: the AP time sequence is split into segments made of 16 consecutive APs frames. Encoding an AP segment includes determining the 16 coefficient values by using DCT, and quantizing and coding separately the DC and AC coefficients, whose prediction errors are coded by using Huffman tables.

Both encoding methods offer good performances when dealing with very low bitrates, e.g., 4 kb/s for the first method and 1.3 kb/s for the DCT-based one, when dealing with a Body Animation Parameter (BAP) stream. Whereas predictive-based coding, using such a frame-based approach, is appropriate for real-time applications, the DCT-based method, using a temporal buffer of 16 frames, is more suitable for off-line applications.

11.5.2.2 The Present: Bone-Based Animation (BBA) and Morphing

BBA [30] is a compression tool for geometric transforms of bones (used in skinning-based animation) and weights (used in morphing animation). These elements are defined in the scene graph and should be uniquely identified. The BBA stream refers to these identifiers and contains, at each frame, the new transforms of bones (expressed as Euler angles or quaternions) and the new weights of the morph targets.

A key point for ensuring a compact representation of the BBA animation parameters consists in decomposing the geometric transformations into elementary motions. For example, when using only the rotation component of the bone geometric transformation, a binary mask indicates that the other components are not involved. The compactness of the animation stream can still be improved when dealing with rotations by expressing them in their respective local coordinate systems. A rotation can be represented either as a quaternion (four fields to be encoded) or as a set of Euler angles (three fields).

BBA follows a traditional signal compression scheme by including two encoding methods as illustrated by Fig. 11.7. Similarly to what was done in the FBA framework, two compression schemes have been developed for encoding the Skeleton, Muscles and Skin (SMS) animation parameters, namely predictive-based and DCT-based.

In the first method, the animation parameters are quantized and coded by a predictive coding scheme. For each parameter to be coded in frame *n*, the decoded value of

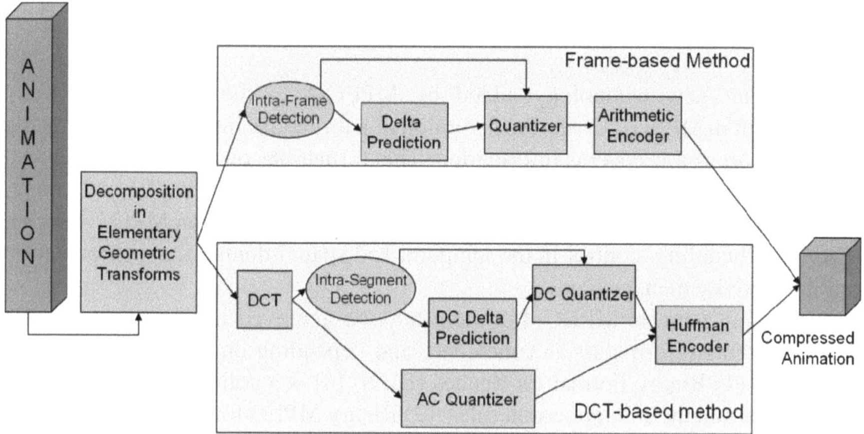

Fig. 11.7 BBA encoder

this parameter in frame *n*-1 is used as a prediction. The prediction error is then encoded using adaptive arithmetic coding. This scheme prevents encoding error accumulation. To control the encoding precision, the quantization step is obtained by multiplying two values, both exposed to the user. The first ensures an exponential control (2^x) and the other a directly proportional one (x). The quantized values are passed to the adaptive arithmetic encoder. The coding efficiency is increased by providing the encoder with range estimates for the animation parameters.

The DCT-based coding method splits the animation parameters time sequences into segments made of 16 consecutive frames. Encoding a segment includes three steps achieved for all the animation parameters: (1) determining the 16 coefficient values by using DCT; (2) quantizing and coding the AC coefficients; and (3) quantizing and differentially coding the DC coefficients. The global quantization step Q for the DC coefficients can be controlled and the global quantization step for AC coefficients is $Q/3$. The DC coefficient of an intra-coded segment is encoded as it is; for an inter-coded segment, the DC coefficient of the previous segment is used as a prediction for the current one. Huffman coding is used for the prediction error of the AC coefficients for both intra-and inter-coded segments.

The compression ratios obtained with both encoding schemes depend strongly on the motion and skeleton complexity, and on the number of muscles used. Typically, when considering a human-like virtual character with natural motion (obtained from a motion capture system), a bit-rate of 25 kb/s ensures a good motion quality, although some optimization mechanisms allow achieving better results. Without an a priori knowledge about the range of bone and muscle motion, due to the fact that any kind of skeleton is allowed, the MPEG-4 standard cannot directly address an efficient arithmetic coding. For this reason, the standard supports the transmission of this range within the animation stream. An optimized implementation [31] of the BBA encoder obtains up to 70:1 compression factor with respect to a textual representation.

11.6 Scene Graph Compression Tools

A significant new technology added by MPEG-4 to the techniques already standardized in MPEG-1/2 was the definition of a format for interactive, multimedia scenes as part of the systems information. This format lies on top of all media data and acts as an intermediate layer between media data and displayed content. It provides a flexible way to manipulate various media types in an MPEG-4 scene, allowing scheduling, control in the temporal and spatial domains, synchronization and interactivity management.

Delivering generic MPEG-4 content may be achieved in three ways, as far as the systems information is concerned, and depending on the application (see Chap. 10): (1) Binary Format for Scenes (BIFS) [8] is a collection of 2D and 3D scene nodes and their compression rules; (2) Binary MPEG format for XML (BiM) consists in applying a generic XML compression tool to XMT, the XML-based, textual description of MPEG-4 scene graph nodes; (3) Lightweight Application Scene Representation (LASeR) is a subset of 2D nodes, designed for building applications targeted at low-performance, mobile terminals. Section 11.6.1 summarizes and discusses the main features of the three approaches.

As for 3DG content, given the penetration of COLLADA in the 3DG market, as well as the adoption of XML for the new version of VRML97, X3D, MPEG decided to develop the 3DG Compression Model (3DGCM) [9] of the MPEG-4 set of specifications, with the goal of offering an architectural model able to accommodate a third party XML-based description of the scene graph with potential binarization tools and with MPEG-4 3DG compression tools. Section 11.6.2 is devoted to 3DGCM.

11.6.1 BInary Format for Scene (BIFS) and Related Tools

As explained above, there are three ways to deliver generic (i.e., not necessarily 3DG-based) MPEG-4 content: BIFS, XMT+BiM and LASeR. Table 11.1 summarizes the characteristics of the two first approaches vs. the third, and emphasizes how

Table 11.1 Characteristics of the three mechanisms for generic MPEG-4 content delivery

	BIFS/XMT+BiM	LASeR
Targeted applications and terminals	Internet, broadcasting, simple games on any terminal	Simple GUIs: menus, EPGs on low-performance terminals
Complexity	Moderate	Low
Graphics objects	2D and 3D	2D
Basic graphics primitives	Rectangle, circle, quadric, text, box, sphere, cylinder, cone	Line, rectangle, circle, ellipse, text
Complex graphics representation tools	Curve2D, IFS, NURBS, [W]SS	Polygon, polyline
Supported elementary streams	Image, video, audio, animation, compressed 3DG	Image, video, audio, animation

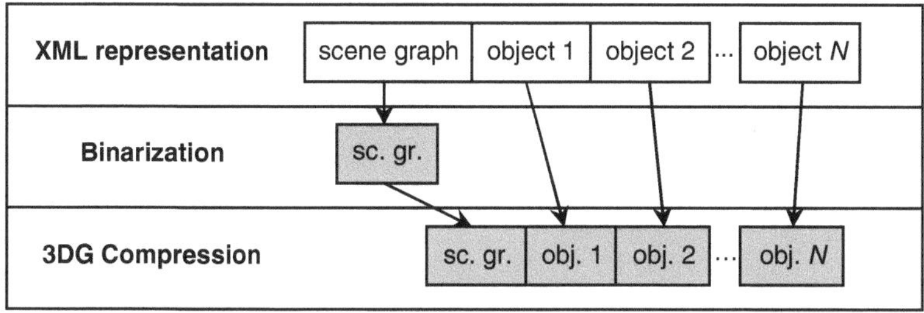

Fig. 11.8 The three layers of the 3DGCM architecture

BIFS and XMT + BiM support both 2DG and 3DG, in contrast to LASeR, which only supports 2DG. In addition, BIFS can accommodate 3DG compression streams. Mainly, BIFS is a binary encoded (so much more compact) version of an extended set of VRML97, with many more 2DG and 3DG primitives. In addition to the coding scheme, consisting in prediction, quantization and entropy coding, BIFS includes the mechanisms to stream the content.

11.6.2 3D Graphics Compression Model (3DGCM)

MPEG-4 Part 25, "3D Graphics Compression Model (3DGCM)" [9], allows to apply compression tools specified in other Parts of MPEG-4 to the elements of a multimedia 3D scene, even if the scene graph chosen by the application designer is not the one defined by MPEG-4. The bitstreams obtained with this architectural model are formatted as MPEG-4 data, and advanced compression features such as progressive decompression and support for streaming, provided by the specific elementary stream compression tools, are preserved, since the decoding of the scene graph is decoupled from that of the object graph.

As shown in Fig. 11.8, the architecture of 3DGCM has three layers:

1. *(Textual) XML representation* layer: The current model can accommodate any scene graph and graphic primitives representation formalism. The only requirement for this representation is that it be expressed in XML. Any XML Schema (specified by MPEG or by external bodies) may be used. Currently the following XML Schemas are supported: XMT, COLLADA and X3D, meaning that correspondence between the XML element and the compression tool handling it is specified in the standard.
2. *Binarization* layer: The binarization layer processes the XML data that is not encoded by the MPEG-4 specific elementary bitstream encoders (e.g. scene graph elements). This data is encapsulated in the "meta" atom of the MP4 file and can be converted to a textual representation or binarized by using gzip (NB: gzip is natively supported by MPEG-4 for encoding the "meta" atom).
3. *3DG Compression* layer: This layer includes the following MPEG-4 tools discussed above: 3DMC, WSS, CI, OI, PI, BBA and FAMC.

The advantage of 3DGCM is that it allows to use the powerful 3DG compression tools in MPEG-4 while keeping the generality of other scene graph and graphic primitives representations.

11.7 Application Examples

One of the straightforward usages of MPEG-4 3DG tools is the rendering of 3D assets within a player, similarly to what is done with audio/video content. Fig. 11.9 shows (of course, static pictures of) animated virtual characters rendered in an MPEG-4-compliant player.

In the European research project FP6-IST-1-507926, OLGA (a unified framework for On-Line GAming), MPEG-4 3DG tools were used for compressing animated and textured 3D objects, and a framework was established to develop distributive multi-player on-line games, or other multimedia applications with heavy and highly variable bandwidth and rendering requirements [32]. The OLGA framework hooks to a complete toolkit of standardized multimedia content compression formats such as MPEG-4 and JPEG 2000, enabling easy deployment over existing infrastructure, while not impeding well-established practices in the game development industry. As a proof of concept, a multi-player on-line game was designed and implemented on two very different kinds of terminals: PCs and cell phones, as shown in Fig. 11.10. The PCs had powerful graphics cards and high bandwidth network connections, whereas the cell phones had no HW acceleration support for 3DG and UMTS connection at best, yet players of both terminals were able to share a common game experience.

The richness of MPEG-4, and especially BIFS, with respect to the graphics primitives, user input (sensors), propagation of actions in the scene graph (routes) and programming application behavior (scripts and Java applets), enables the programming of complete 2D/3D games and interactive applications to be interpreted by a compliant player or browser, as illustrated by Fig. 11.11.

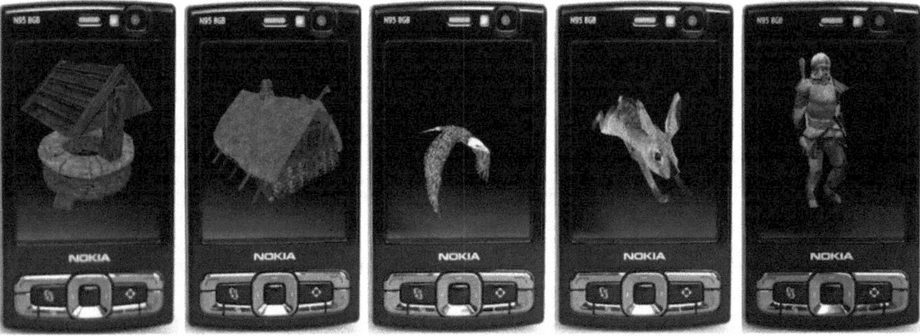

Fig 11.9 3DG assets visualized on a cell phone implementation of an MPEG-4 player

Fig. 11.10 Screen shots from both the PC and cell phone versions of OLGA's multi-platform on-line game with inherently scalable 3D content

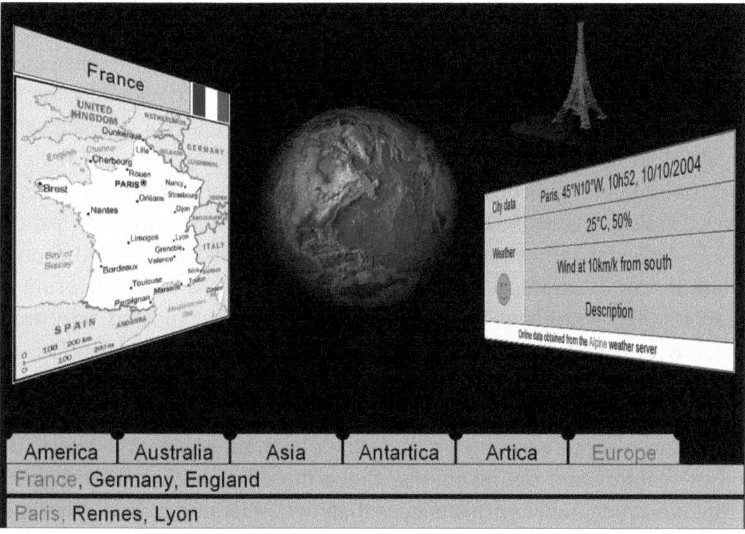

Fig. 11.11 A geographic encyclopedia visualized by an MPEG-4 media player

The complexity of using scripts or applets embedded in the content itself makes it suitable for powerful platforms but less appropriate for mobile phones. Other than the traditional approach of interpreting locally the 3DG primitives and computing the pixel maps, the "intrinsically on-line" nature of mobile phone allows the alternative approaches of remote computing and/or rendering. The later consists in interpreting the graphics primitives on a server (much more powerful than the mobile phone), and then send the result as a video stream to the mobile device [33]. In the upper channel, the user interaction with the terminal (keyboard and screen touch) is directly transmitted to the server. The only processing requirement on the client side is to have a terminal able to decode and display the video stream. The major limitation of this solution is the need of significant bandwidth, making it feasible only when the network connection can insure good transmission quality. On the server side this approach requires significant computational power, and leads to poor performances if too many users are using the system simultaneously. Additionally, the graphics card(s) of the server must also be shared between the users. As an alternative, the rendering can be distributed on dedicated rendering clusters, thus supporting more clients.

Another idea is to execute the game logic on the server and the rendering on the mobile terminal. In addition, to ensure that, during the play game, there are no situations with high peaks of data transmission, the entire initial scene and media layers can be encapsulated and transmitted before starting the game. This operation is performed by using the MPEG-4 file format, able to represent complex 3D graphics scenes with associated multimedia data. Furthermore, it is possible to adapt the graphics content for a specific terminal [34] allowing for the best possible trade-off between performance and quality. By building the architecture on top of the MPEG-4 standard, the communication between the rendering engine and the game logic follows a standardized formalism. This allows to create them separately, being therefore possible for developers to generate game logic for different games that will use the same rendering engine, without taking into account how the latter is implemented. On the other hand, the developers of the rendering engine can create optimized solutions for a specific platform without knowing the game logic. In this case, the communication is based on a higher level control mechanism: the graphics primitives can be grouped and controlled as a whole by using only a few parameters. The MPEG-4 standard allows any part of the scene to be loaded, changed, reorganized or deleted. As an example, the same game can be played on a rendering engine that supports 3D primitives, probably accelerated by dedicated hardware, and simultaneously on a rendering engine that only supports 2D primitives. This flexible approach allows distribution of the games on multiple platforms without the need to change the code of the game logic. Another advantage is the possibility to improve the game logic without additional cost to the client, allowing easy bug-fixing and addition of features and optimizations. Since the game logic can be hosted on a server that has much better performances than the client terminal, it can implement more advanced features that normally will not be available on the client terminal. These features may include artificial intelligence, advanced physics and any other techniques usually requiring a large computational power. These improvements will not change anything on the user side, so the user will be able to play more complex games on the same terminal. A snapshot of a game built with this architecture in mind is presented in Fig. 11.12.

Fig. 11.12 Snapshot of a race game visualized with the MPEG-4 player and using a remote server for computing the game logic

The flexibility of the proposed architecture makes it appropriate for powerful terminals as well. In these conditions, it is possible to download the game logic server software, install it on the terminal and use it to play the game locally. However, to keep this approach effective, the games should be designed from the start having in mind the requirements and the restrictions. The rendering engine can also be integrated in the firmware of the device, allowing the manufacturer to optimize it for the specific hardware. Even more, the different decoders supported by MPEG-4 (video, audio, 3DG, animation) can be integrated in a dedicated hardware to improve performance.

References

1. I. E. Sutherland, "Sketchpad: A Man-Machine Graphical Communication System", MIT Ph. D. thesis, January 1963.
2. Web3D Consortium and ISO/IEC JTC 1/SC 24, "ISO/IEC 19775-1:2008, Information technology – Computer graphics and image processing – Extensible 3D (X3D), Part 1: Architecture and base components", ISO/IEC International Standard (IS), July 2008. This is the 2nd and most recent Edition of this IS, including all Corrigenda and Amendments to its previous Edition, which was published by ISO in November 2004.
3. Virtual Reality Modeling Language (VRML) Consortium (now Web3D Consortium) and ISO/IEC JTC 1/SC 24: "ISO/IEC 14772-1:1997, Information technology – Computer graphics and image processing – The Virtual Reality Modeling Language – Part 1: Functional specification and UTF-8 encoding", ISO/IEC IS, December 1997. This is the 1st and most recent Edition of this IS, commonly known as "VRML97".
4. R. Arnaud and M. C. Barnes, "COLLADA: Sailing the Gulf of 3D Digital Content Creation", A. K. Peters, Wellesley, 2006.

5. Moving Picture Experts Group (MPEG; formally ISO/IEC JTC 1/SC 29/WG 11): "ISO/IEC 14496-16:2009, Information technology – Coding of audio-visual objects – Part 16: Animation Framework eXtension (AFX)", ISO/IEC IS, December 2009. This is the 3rd and most recent Edition of this IS, including all Corrigenda and Amendments to its previous Editions, of which the 1st was published by ISO in February 2004 (the technical committee work and the balloting process had finished in December 2002). The 4th Edition has also been finalized and should be published by ISO within 2011.
6. J. D. Foley, A. van Dam, S. K. Feiner and J. F. Hughes: "Computer Graphics: Principle and Practice (2nd ed. in C)", Addison-Wesley, 1997.
7. MPEG: "ISO/IEC 14496-2:2004, Information technology – Coding of audio-visual objects – Part 2: Visual", ISO/IEC IS, May 2004. This is the 3rd and most recent Edition of this IS, including all Corrigenda and Amendments to its previous Editions, of which the 1st was published by ISO at the beginning of 1999.
8. MPEG: "ISO/IEC 14496-11:2005, Information technology – Coding of audio-visual objects – Part 11, Scene description and application engine", ISO/IEC IS, December 2005. This is the 1st and most recent Edition of this IS.
9. MPEG: "ISO/IEC 14496-25:2009, Information technology – Coding of audio-visual objects – Part 25, 3D Graphics Compression Model", ISO/IEC IS, March 2009. This is the 1st and most recent Edition of this IS.
10. MPEG: "ISO/IEC 14496-5:2001, Information technology – Coding of audio-visual objects – Part 5, Reference software", ISO/IEC IS, December 2001. This is the 2nd and most recent Edition of this IS, including all Corrigenda and Amendments to its previous Edition, which was published by ISO in May 1999.
11. MPEG: "ISO/IEC 14496-27:2009, Information technology – Coding of audio-visual objects – Part 27, 3D Graphics conformance", ISO/IEC IS, December 2009. This is the 1st and most recent Edition of this IS.
12. G. Farin: "Curves and Surfaces for CAGD: A Practical Guide (5th ed.)", Morgan Kauffman, 2001.
13. M. Avilés and F. Morán: "Static 3D triangle mesh compression overview", Proc. IEEE Intl. Conf. Image Processing (ICIP) 2008, p. 2684–2687, October 2008.
14. G. Taubin and J. Rossignac: "Geometric Compression Through Topological Surgery", ACM Trans. Graphics, vol. 17, nr. 2, p. 84–115, April 1998.
15. E.-Y. Chang, N. Hur and E. S. Jang: "3D model compression in MPEG", Proc. IEEE ICIP 2008, p. 2692–2695, October 2008.
16. H. Hoppe: "Progressive Meshes", Proc. ACM SIGGRAPH 1996, p. 99–108, August 1996.
17. G. Taubin, A. Guéziec, W. Horn and F. Lazarus: "Progressive Forest Split Compression", Proc. ACM SIGGRAPH 1998, p. 123–132, July 1998.
18. A. Khodakovsky, P. Schröder and W. Sweldens: "Progressive Geometry Compression", Proc. ACM SIGGRAPH 2000, p. 271–278, July 2000.
19. F. Morán and N. García: "Comparison of Wavelet-Based Three-Dimensional Model Coding Techniques", IEEE Trans. Circuits and Systems for Video Technology, vol. 14, nr. 7 (special issue on AFX), p. 937–949, July 2004.
20. I. A. Salomie, A. Munteanu, A. Gavrilescu, G. Lafruit et al.: "MeshGrid – A Compact, Multiscalable and Animation-Friendly Surface Representation", IEEE Trans. Circuits and Systems for Video Technology, vol. 14, nr. 7, p. 950–966, July 2004.
21. J. Royan, R. Balter and C. Bouville: "Hierarchical Representation of Virtual Cities for Progressive Transmission over Networks", Proc. Intl. Symp. 3D Data Processing, Visualization and Transmission (3DPVT) 2006, p. 432–439, June 2006.
22. Joint Photographic Experts Group (JPEG; formally ISO/IEC JTC 1/SC 29/WG 1): "ISO/IEC 10918-1:1994, Information technology – Digital compression and coding of continuous-tone still images: Requirements and guidelines", ISO/IEC IS, February 1994. This is the 1st and most recent Edition of this IS, commonly known as "JPEG".

23. JPEG: "ISO/IEC 15444-1:2004, Information technology – JPEG 2000 image coding system: Core coding system", ISO/IEC IS, September 2004. This is the 2nd and most recent Edition of this IS, including all Corrigenda and Amendments to its previous Edition, which was published by ISO in December 2000.
24. G. Lafruit, E. Delfosse, R. Osorio, W. van Raemdonck et al.: "View-Dependent, Scalable Texture Streaming in 3D QoS with MPEG-4 Visual Texture Coding", IEEE Trans. Circuits and Systems for Video Technology, vol. 14, nr. 7, p. 1021–1031, July 2004.
25. F. Morán (ed.): "Efficient representation of 3D meshes with multiple attributes", Proposed Draft AMendment (PDAM) of Amendment 1 to MPEG-4 Part 16 Edition 4, MPEG private output document N11456, July 2010.
26. E. S. Jang, J. D. K. Kim, S. Y. Jung, M.-J. Han et al.: "Interpolator data compression for MPEG-4 animation", IEEE Trans. Circuits and Systems for Video Technology, vol. 14, nr. 7, p. 989–1008, July 2004.
27. K. Mamou, T. Zaharia and F. Prêteux: "A skinning approach for dynamic 3D mesh compression", Computer Animation and Virtual Worlds, vol. 17, nr. 3–4, p. 337–346, July 2006.
28. I.-S. Pandzic, N. Magnenat Thalmann, T. K. Capin and D. Thalmann: "Virtual Life Network: a Body-Centered Networked Virtual Environment", Presence, MIT, vol. 6, nr. 6, p. 676–686, December 1997.
29. T. K. Capin, I.-S. Pandzic, N. Magnenat Thalmann and D. Thalmann: "Virtual Human Representation and Communication in VLNet", IEEE Computer Graphics and Applications, vol. 17, nr. 2, p. 42–53, March 1997.
30. M. Preda, I. A. Salomie, F. Prêteux and G. Lafruit: "Virtual character definition and animation within the MPEG-4 standard", p. 27–69 in M. Strintzis and N. Sarris (eds.): "3D modeling and animation: Synthesis and analysis techniques for the human body", IRM Press, 2004.
31. M. Preda, B. Jovanova, I. Arsov and F. Prêteux: "Optimized MPEG-4 animation encoder for motion capture data", Proc. ACM Intl. Conf. 3D Web Technology (Web3D), p. 181–190, April 2007.
32. F. Morán, M. Preda, G. Lafruit, P. Villegas and R.-P. Berretty: "3D Game Content Distributed Adaptation in Heterogeneous Environments", EURASIP Journal on Advances in Signal Processing, vol. 2007, art. 93027 (15 p.), September 2007.
33. I. Nave, H. David, A. Shani, A. Laikari et al.: "Games@Large graphics streaming architecture", Proc. IEEE Intl. Symp. Consumer Electronics (ISCE), p. 1–4, April 2008.
34. S. M. Tran, M. Preda, F. Prêteux and K. Fazekas: "Exploring MPEG-4 BIFS features for creating multimedia games", Proc. IEEE Intl. Conf. Multimedia and Expo (ICME), vol. 2, p. 429–432, July 2003.

Chapter 12
MPEG Reconfigurable Video Representation

Marco Mattavelli

12.1 Introduction

12.1.1 Background and Objectives of RVC

The standardization efforts of MPEG in video coding, originally had as main objective to guarantee interoperability of compression systems. This carried with it the possibility to reach another important objective, namely to wide and easy deployment of implementations of those standards. While at the beginning MPEG-1 and MPEG-2 were only specified by textual descriptions, with the increasing complexity of video coding tools, starting with the MPEG-4 set of standards, C or C++ specifications, called *reference software*, have also become a formal specification of the standards. However, descriptions composed of non-optimized non-modular software packages have shown limitations. Since in practice they are frequently the starting point of an implementation, system designers must rewrite these software packages not only to try to optimize performance, but also to transform such specifications into appropriate forms adapted to be the starting point of current system design flows.

In addition the evolution of video coding technologies leads to solutions that are increasingly complex to implement and present significant overlap between successive standards. Traditionally, the monolithic nature of the reference software required complete rewrites of the code for new standards and failed to take advantage of the overlap of technologies, effectively obscuring what is new and what is common between an old and new standard specification.

Another problem of traditional monolithic sequential specifications is that they hide the inherent data flow structure of the video coding algorithms, which is a

M. Mattavelli (✉)
Ècole Polytechnique Fédérale De Lausanne, CH-1015, Lausanne, Switzerland
e-mail: marco.mattavelli@epfl.ch

L. Chiariglione (ed.), *The MPEG Representation of Digital Media*,
DOI 10.1007/978-1-4419-6184-6_12, © Springer Science+Business Media, LLC 2012

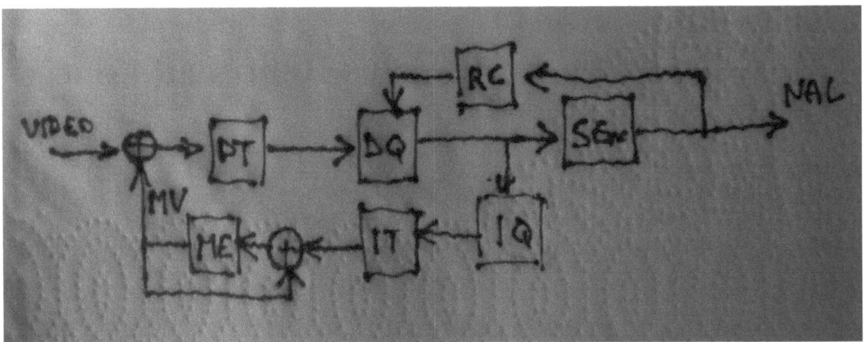

Fig. 12.1 Example of the specification of a video coding algorithm that can be obtained by a video engineer during lunch-time. As can be noticed, it is in the form of "boxes" that represent processing units and arrows that represent flows of data. No control loops of sequential programming languages or abstract objects, classes or methods are used[1]

significant impediment for efficient implementation on any class of target platforms being video coding a *data dominated* algorithm. The intrinsic concurrency and parallelism of any modern video coding tool is completely obscured by such reference software written using pure sequential paradigm and languages. The required analysis and rewriting of the specification is difficult and time consuming, as it involves transforming a highly complex sequential algorithm relying heavily on global storage into appropriate implementations that exploit the manifold forms of concurrency in it, and whose observable behavior is identical. While in the past such problem was mainly faced by hardware system designers, with the appearance of the new generation of multicore processors has now become a challenging problem that also software designers need to tackle (Fig. 12.1).

Another problem originated by the past MPEG specification form, and consequence of the wide variety of video coding algorithms after almost 20 years of MPEG activity, is the selection of *profiles*. The a priori specification of a small number of such profiles has become very problematic. Since some video tools are rather complex and apt only to be implemented on certain devices, or are not required for all applications, they are included only into standardized *profiles*. Interoperability is thus guaranteed at the level of these standard profiles. However, such trade-off prevents the possibility of optimally satisfying a variety of specific applications, whereas the specification of too many profiles would result in an obstacle to interoperability. In any case the lack of modularity and encapsulation properties of the past specification form does not help to support more flexible and dynamic ways to select such algorithm subsets to satisfy other specific application requirements. Those requirements could also include anything besides coding efficiency, which is the traditional focus of current standardization efforts. Although coding efficiency is of fundamental importance, other trade-offs, such as coding performance (in terms of overall computational effort or latency), are not efficiently realizable staying within current

[1] Picture with the kind permission of Jorn Janneck.

standard profiles. Moreover, the possibility of seeking these new trade-offs for codecs and the motivations that lies behind such search of new profiles are increased on the one hand by the large number of coding algorithms available in existing standard technology (MPEG-1, MPEG-2, MPEG-4 Part 2 and part 10 including AVC and SVC) and on the other hand by the increasing number of application domains employing video compression, in addition to the historical digital video broadcasting and storage, which are looking only for very specific optimizations.

Last, but not least, the current scheme for the definition and standardization of new video coding technology results in a noticeably long time span between a new concept is validated until it is implemented in products and applications as part of a worldwide standard. Each MPEG standard can be considered as a frozen version or a snapshot of state-of-the-art of video compression taken a few years before the standard is released in its final form to the public. This delay is a significant impediment to providing innovations to the market, thus opening the way to ad-hoc proprietary solutions that jeopardize all efforts made by standardization bodies to ensure interoperability and openness to state of the art technology to all operators.

Considering all these problems and drawbacks of the process used so far for developing a new standard, MPEG has finalized a new standard framework called Reconfigurable Video Coding (RVC). The goal of the MPEG RVC effort is to address the limitations and problems discussed above and thus offer a more flexible use and faster path to innovation of MPEG standards in a way that is competitive in the current dynamic environment [1–3].

12.1.2 New Concepts and Formalisms of the RVC Specification

A paradigm shift, for both specification formalism and for the concept of a standard video codec itself, is necessary in order to achieve these objectives. The basic notion of the new RVC standard is the coded representation of the decoder architecture based on a standard library of video coding algorithms. Whereas so far a standard decoder is defined a priori and can only decode bitstreams conformant with such predefined standard, the MPEG RVC framework supports a new different scenario.

As illustrated in Fig. 12.2, a decoder is defined by a decoder description (a decoder configuration plus a bitstream syntax), not known a priori, and such information is sufficient to instantiate a decoder capable of decoding new conformant bitreams of encoded video content. The information that is known a priori in the new MPEG RVC standard is the unified library of video coding algorithms. Therefore, what becomes standard is the *module* of a library and the formalism used to specify a decoder architecture. Such *module* is the basic *brick* for building decoders or better new decoder configurations. Libraries can be updated incrementally as soon as new video technology is demonstrated to be valuable for applications simply by adding new *modules.*

However, to ensure that a collection of *modules* can be a rigorous form of specification, without any ambiguity in the way the decoding process is specified, specific constraints need to be satisfied. In particular it is required to satisfy *composability* properties, i.e. to guarantee that any composition of building blocks constitutes a

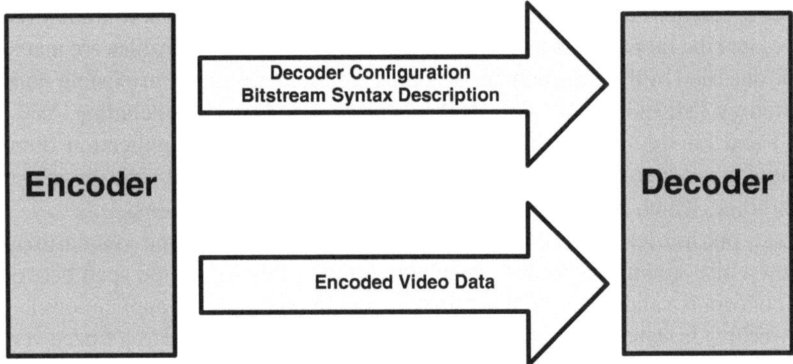

Fig. 12.2 Conceptual view of the RVC framework. A decoder capable of decoding a given encoded video data is fully specified by a decoder configuration and by the syntax of the video data bitstream

specification of a specific decoder behavior without any ambiguities for generating compliant implementations [4]. Building a video codec specification by composing such independent modules that *encapsulate* in their I/O behavior the essential part of the video coding technology is very attractive and powerful. However, for doing that it has been necessary to abandon the sequential specification paradigm (i.e. the Von Neumann machine paradigm that constitute the base of most of the commonly used programming languages such as C, C++, Java) and to adopt a different *data flow* based computation model and language [5, 6]. Among the possible computation models supporting the concept of data packets flowing throughout a configuration (i.e. network) of Functional Units the one so called *asynchronous data flow* was chosen as the most general and capable of specifying the widest class of algorithms. Other computational models such as *synchronous data flow* result simpler to analyze and to implement, but are not capable of implementing dynamic data processing typical of video coding such as for instance parsing a typical MPEG bitstream [7]. The language and associated computation model that was found to better support the desired properties and requirements identified by MPEG was the actor dataflow language called CAL originally designed by J. Janneck and J. Eker [5]. Since such language was not a standard, a profile of such language called RVC-CAL has been expressly specified and standardized in ISO/IEC 23001-4. Thus an RVC codec is specified as a *configuration* of *standard modules* called in RVC Functional Units (FU) and only characterized by their input–output behavior specified using RVC-CAL. Such configuration is a data flow program that can be seen as a directed graph, where the nodes represent the units in which computation can occur (Functional Units in RVC), and the arcs represent the flow of data, or more rigorously, a token stream flowing along each arc. Functional Units have input and output ports through which they receive and send tokens from and to their environment. Each connecting arc leads from an output port to an input port. The flow of tokens along those arcs is lossless and order preserving. This implies that every token that the sending FU emits at the output port is guaranteed to arrive at the connected input port of another FU, or the same FU in the case of self-loop. Tokens arrive in exactly the order in which they were sent. No other assumptions are made beyond this, in particular not with respect to timing, from which the asynchronous data flow model abstracts from. FUs are loosely coupled and consequently, FUs send and receive tokens in relative asynchrony, and tokens may be buffered on the way from the sender to the receiver.

Thus, any implementation, in whatsoever form, that respects these quite intuitive and simple dataflow and computation assumptions is guaranteed to have an external observable behavior identical to the RVC specification from which it has been based or derived.

This important property on one hand gives the foundation of the conformance point of RVC specifications, and on the other leaves the widest freedom to designers to implement the specification in the form that is more appropriate to the specific target platform and application requirements.

More formally we can define the model of computation employed in RVC as follows [4, 5]:

1. An RVC Abstract Decoder Model (ADM) or more simply an RVC decoder specification is a dataflow program composed of a static network of concurrently operating entities Functional Units (FUs) connected by lossless, order-preserving, directed channels over which they exchange data objects (*tokens*).
2. Functional Units can have state, and for any two Functional Units their state does not overlap.
3. Each of those Functional Units operates by performing a sequence of terminating, uninterruptible (*atomic*) steps. During each step, it can
    (a) Consume input tokens,
    (b) Produce output tokens,
    (c) Modify its state.
4. The conditions that guard the execution of a step may be any predicate that involves the presence of input, the value of input required to be present, and the state of the Functional Unit.
5. Different steps may consume and produce different numbers of tokens.

A corollary of (3)'s requirement that steps be terminating is that all conditions for termination must be testable at the outset, i.e. before a step is started (such as, for instance, the presence of all input tokens required for the step), and those conditions must be guarding a step. A corollary of (2)'s non-overlapping state requirement is that the only way in which two Functional Units may communicate is by exchanging tokens with each other using an asynchronous data flow computation model.

As mentioned earlier each FU module behavior is specified in RVC-CAL and the connections between FUs output and input ports by an XML based language called in RVC Functional unit Network Language (FNL). Both these two languages have been expressly standardized by MPEG in RVC with the objective of, on one side, to provide all functionality and expressiveness appropriate for the RVC specification goals (modularity, compactness, expressiveness), and on the other side to present a consistent basis for the development of efficient tools that can directly synthesize both software and hardware implementations. In fact the new RVC specification formalisms not only provides advantages, even when employing the classical methods based on hand writing software code or hardware description language (HDL) code, but also provides the possibility of generating by direct synthesis different forms of implementations such as single processor SW, multi-core SW, HDL and any combination of thereof [8–16], which is certainly a much more attractive option.

In summary, the MPEG RVC effort has the ambition to provide a more flexible way of providing standard video technology using a specification formalism laying at a higher level of abstraction than the one traditionally used, but at the same time provide a starting point for implementation that is compatible with more efficient methodologies for developing implementation for the emerging and future generation of processing platforms [2, 3, 17–19].

12.1.3 The Structure of the Standard: MPEG-B Part 4 and MPEG-C Part-4

A pictorial view of the components of the RVC framework is reported in Fig. 12.3. The upper part represents the standard components that are used to build an RVC specification. On the top left the *Decoder Description* is composed by the configuration of FUs expressed in FNL plus the description of the bitstream syntax. The configuration of FUs plus the specification of their behavior available from the standard MPEG Tool Library constitute the so called *Abstract Decoder Model* (ADM) that is the executable specification of a decoder in the form of a data flow program satisfying the asynchronous model of computation. The syntax of the bitstream that an ADM need to be able to parse is expressed using a BSDL (Bitstream Syntax Description Language) description. BSDL is a XML-based language that specifies the sequence of the syntax elements with possible conditioning on the presence of the elements, according to the value of previously decoded elements. BSDL is explained in more details in 10.2.2.2. All upper section of the picture represent standard components of RVC. The bottom part represent the non normative implementation of an RVC decoder specification showing how decoders can be derived from a given ADM by using direct synthesis and/or by instantiating any proprietary implementations of FUs that have an identical internal behavior of the FU specified by the RVC standard.

Two standards are defined so far in the MPEG RVC framework: ISO/IEC 23001-4 (also called MPEG-B part 4) and ISO/IEC 23002-4 (MPEG-C part 4).

ISO/IEC 23001-4 defines the overall framework as well as the standard languages that are used to specify a new codec configuration of an RVC decoder. The *Abstract Decoder Model* (ADM) is an executable description using the modular data flow

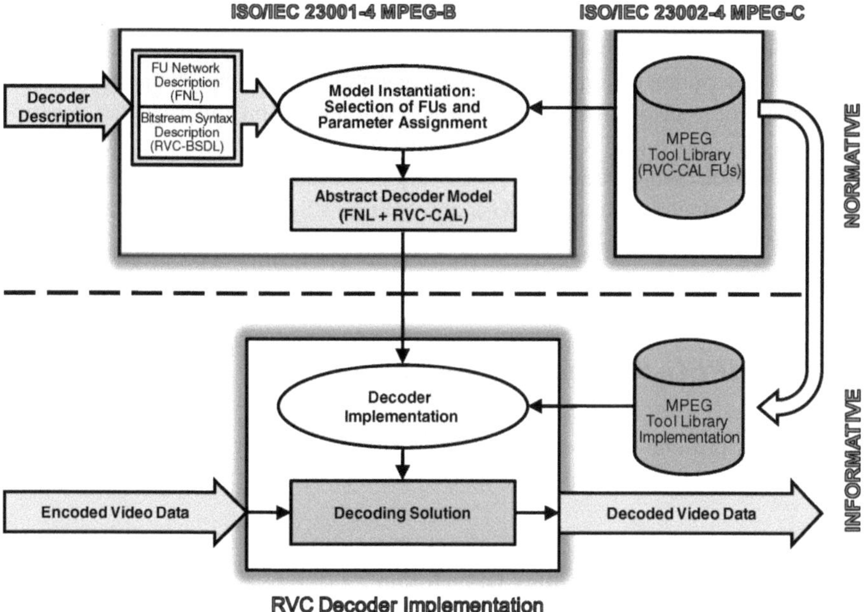

Fig. 12.3 The normative and non-normative components of the RVC framework. The normative components are the standard languages used to specify the ADM and the standard library of FU. The informative components are a schematic illustration of how implementations can be derived by any methodology including tools that directly synthesize a decoder implementation possibly using proprietary implementations of the standard library

model and constitutes the essential component of the specification of a new codec configuration. In more details the MPEG-B standard includes:

- The specification of the Functional unit Network Language (FNL), which is the language describing the video codec configurations. FNL is an XML-based language that provides the instantiation of the Functional Units (FUs) composing the codec, their parameterization, as well as the specification of the connections.
- The specification of the RVC-Bitstream Syntax Description Language (RVC-BSDL), which is a subset of the standard MPEG BSDL (fully specified in ISO/IEC 23001–5); a language syntactically describing the structure of the input encoded bitstream.
- The specification of the RVC-CAL, the dataflow language that is used to express the behavior of each FU and consequently the behavior of any network of FUs described in FNL. RVC-CAL is used to specify all MPEG FUs that compose the *RVC MPEG Toolbox* specified in ISO/IEC 23002-4.

ISO/IEC 23002-4 specifies the unified library of video coding algorithms employed in the current MPEG standards. Up to now, several MPEG standards/profiles are fully covered: MPEG-4 part 2 Simple Profile, MPEG-4 part 10 (AVC) Constrained Baseline Profile, Progressive High profile and Scalable Baseline profile, but the plans are to cover all standards for the most used MPEG profiles such as MPEG-4 Advanced Simple and MPEG-2 Main profile.

Beside the specification of decoders using MPEG-B and MPEG-C the RVC framework can support more dynamic specifications of video codecs by associating a decoder description to compressed video content. Investigations on the mechanisms supporting the transport of RVC codec description and bitstream syntax descriptions are currently under way. Various scenarios enabling downloads and dynamic update of codec configurations on processing platforms are analyzed so as to verify what (if any) amendment to MPEG-2 and MPEG-4 Systems standards are needed to support the widest class of deployment scenarios for RVC codecs.

12.1.4 MPEG-B Standard Components

This section presents in more details and with some simple examples the various components and languages that are used to build an RVC decoder specification.

12.1.4.1 The Bitstream Syntax Description

The RVC framework aims also at supporting the development of new decoder configurations based on MPEG standard technology, and this implies that the bitstream syntax may change from a decoding solution to another. The consequence is that a new parser needs to be generated for each new bitstream syntax. The parser FU may be the most complex FU in the MPEG toolbox if we take the examples of the existing MPEG standards such as MPEG-4 SP and AVC. A specific MPEG technology capable of specifying in a formal and compact form the syntax of a bitstream exists. It has been adopted in the RVC framework to provide the specification of the parsing process of a new or existing decoder configuration.

12.1.4.2 The RVC-BSDL Language

ISO/IEC 23001-5 is the standard that specifies the so called Bitstream Syntax Description Language (BSDL), an XML-based language used to specify generic bitstream syntaxes. A Bitstream Syntax Description Language (BSDL) schema represents in a compact form all the possible bitstream structures that conforms to such specific syntax description [20]. A single instance of a bistream is called BSD (Binary Syntax Description) and is a unique instance among all possible instantiations of a BSDL description. It is not constituted by a BSDL schema, but by a XML file containing the data of the bitstream. Figure 12.4 (top) shows an example of BSDL schema associated to one possible corresponding BS instance shown in Fig. 12.4 (bottom).

A video bitstream is composed by a sequence of binary elements of the syntax having different lengths. Some elements are composed by a single bit, whereas others

```
<element name= "NALUnit"
   bs2:ifNext= "00000001">
<xsd:sequence>
   <xsd:element name="startCode" type="avc:hex4" fixed="00000001"/>
   <xsd:element name="nalUnit" type="avc:NALUnitType"/>
   <xsd:element name="payload"/>
</xsd:sequence>
<!- Type of NALUnitType  -->
<xsd:complexType name="NALUnitType">
<xsd:sequence>
   <xsd:element name="forbidden_zero_bit" type="bs1:b1" fixed="0"/>
   <xsd:element name="nal_ref_idc" type="bs1:b2"/>
   <xsd:element name="nal_unit_type" type="bs1:b5"/>
</xsd:sequence>
</xsd:complexType>

<xsd:element name="payload" type="bs1:byteRange"/>
```

```
<NALUnit>
   <startCode>00000001</startCode>
   <forbidden0bit>0</forbidden0bit>
   <nalReference>3</nalReference>
   <nalUnitType>20</nalUnitType>
   <payload>5 100</startCode>
</NalUnit>
<NALUnit>
<startCode>00000001</startCode>
<!-- and so on … -->
</NALUnit>
```

Fig. 12.4 *Top*, example of BS schema, fragment of MPEG-4 AVC syntax. *Bottom*, BSD syntax description of an instance of a MPEG-4 AVC bitstream fragment

may contain several bits. The Bitstream Schema (in BSDL) indicates the length of such binary elements in a human and machine-readable format (hexadecimal, integers, strings …). For example, hexadecimal values are used for start codes as shown in Fig. 12.4.

The XML based formalism allows organizing the description of the bitstream in a hierarchical structure. The Bitstream Schema (in BSDL) can be specified at different hierarchy levels.

BSDL was originally conceived and designed to enable adaptation of scalable multimedia contents in a format-independent manner [20]. Conversely in the RVC framework, RVC-BSDL is used to fully specify and describe video bitstream syntaxes. Thus, in RVC, schemas must specify all the elements of the bitstream syntax, from the top hierarchy level down to the lowest level of hierarchy of syntax elements [21].

Before generating new parsers from bitstream syntax schemas, it must be guaranteed that the schemas are correct, i.e. that the schema correctly reflects the

```
<Instance  id = "FU A">
     <Class name = "Algo_Example1" />
</Instance>
<Instance id = "FU B">
     <Class name = "Algo_Example2" />
</Instance>
```

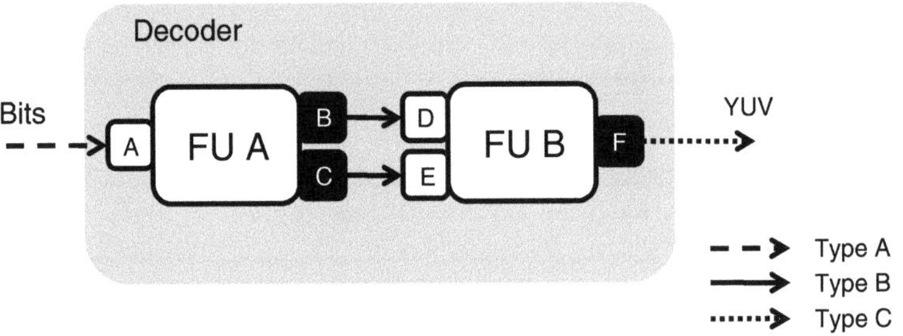

Fig. 12.5 *Top*, example of FND code fragment and corresponding network (*bottom*) showing the three types of possible connections in an FU network

structure of the compressed data which is sent in each instance. Systematic validation procedures are necessary to guarantee that newly generated parsers can decode any instance of a given syntax. Description of such non MPEG normative validation procedures and related tools including the validation of variable length syntax components such as VLD, CAVLD and CABAC [7], are omitted here for brevity. the interested reader can find more details in Ref. [22] and related references.

Once the schema has been validated a parser need to be generated because for a new RVC configuration the corresponding parser might not be available in the standard toolbox library. Since a formal and complete description of the syntax is available in RVC, systematic procedures for the automatic generation of parsers are conceivable. An example of a tool for the generation of an executable parser based on XSL transformations from RVC-BSDL into CALML, an XML-based representation of RVC-CAL, is available in Refs. [22, 23].

12.1.4.3 The FU Network Description Language (FNL)

The Functional Unit Network Language (FNL) is the language that is used to describe networks of FUs, or in other words the configuration of connection arcs among output and input ports of the FUs instantiated from the standard RVC library for a given decoder configuration. FNL is an XML-based language standardized in ISO/IEC 23001-4.

A network of FUs is called a Functional Unit Description (FND) and is specified using FNL. A FND includes the list of the FUs instantiated from the RVC library to form the decoder configuration and the three types of admissible connections, connections between FUs, connections between decoder inputs and FU inputs, connections between FU outputs and decoder outputs, as shown as example in Fig. 12.5.

```
<XDF name="Decoder">
<Instance id = "Syntax parser">
    <Class name = "syntax parser">
</Instance>
<Instance id = "FU A">
    <Class name = "Algo_ExamFU_A">
</Instance>
<Instance id = "FU B">
    <Class name = " Algo_ExamFU_B">
</Instance>
<Instance id = "FU C">
    <Class name = " Algo_ExamFU_C">
</Instance>
<Input src = "Syntax Parser" src-port = "A" />
<Output src = "FU C" src-port = "R" />
<Connection src = "Syntax Parser" src-port = "B" dst = "FU A" dst-
port = "E" />
<Connection src = "Syntax Parser" src-port = "C" dst = "FU A" dst-
port = "F" />
<Connection src = "Syntax Parser" src-port = "D" dst = "FU B" dst-
port = "K" />
<Connection src = "FU A" src-port = "H" dst = "FU C" dst-port = "O"
/>
<Connection src = "FU B" src-port = "L" dst = "FU C" dst-port = "P"
/>
<Connection src = "FU B" src-port = "M" dst = "FU C" dst-port = "Q"
/>
</XDF>
```

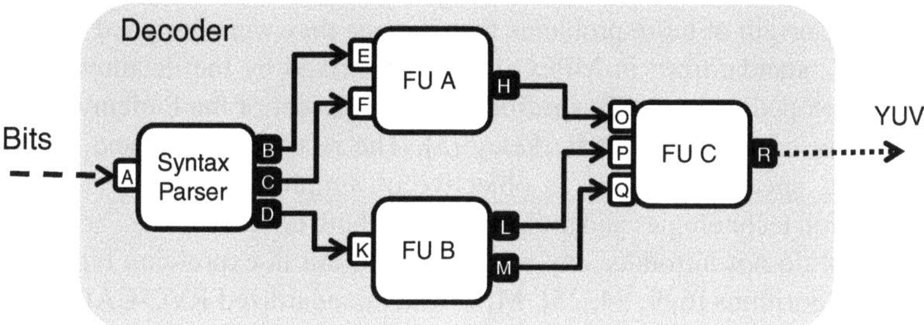

Fig. 12.6 Example of FU network

Unique IDs and names of the FUs instantiated in a given configuration are specified in ISO/IEC 23002-4.

Another example of FU network with four FUs is illustrated in Fig. 12.6.

12.1.4.4 The RVC-CAL Language

The choice made by the selection process of MPEG for the language providing the support of the RVC specification in the form of the Abstract Decoder Model and the reference SW of the RVC Video Tool Library, was to use a subset of a more general dataflow language [5], RVC-CAL, subset that is better designed to the specific

RVC requirements. What has to be underlined is that such choice was quite natural because any program written in RVC-CAL by construction complies with the "RVC model of computation" introduced and discussed in 10.1.2. In fact the operators of the language themselves guarantee that any program written using them complies with the model of computation. Each FUs is specified by a `.cal` file in which input and output ports are defined and the complete network of FUs is specified by a FNL file as described in the previous section.

This fact that is fundamental for setting a *conformance point* for the behavior of a RVC specification (thus on all implementations that conform with it), and could not be guaranteed by using whatever other language such as C or C++ or Java alone. All programs written in such sequential languages will certainly be conformant with the Von Neumann computation model (for which they were designed and used). To respect a different *model of computation* programs not only would need further additional restrictions and constraints when they are written, but in addition they would require specific procedures and appropriate tools to verify that the written program is indeed compliant with the RVC model of computation. One way of doing that could be to develop appropriate normative tools that analyze and verify that a program is compliant with the RVC computation model. However, this would only solve a part of the problem. Further restrictions should be imposed to the usage of operators to make the resulting program analyzable to be able to yield optimized implementations [17]. Such usage of an imperative language restricted by normative tools would result quite cumbersome, unnatural and could have been the origin of more problems than the one they were designed to solve.

RVC-CAL standardized in MPEG has been derived by the dataflow language that was developed and initially specified as a subproject of the Ptolemy project at the University of California at Berkeley [5]. The restrictions adopted versus the original CAL specification had as objective of simplifying the development of implementation technologies and tools supporting both direct hardware and software synthesis, but do not introduce any relevant limitation in expressing typical video processing algorithms [6, 9, 24, 25]. MPEG has standardized RVC-CAL in Annex C of ISO/IEC 23001-4.

RVC-CAL specifies the behavior of FU as basic computational entities (called "actors" in dataflow literature) of a RVC specification. FUs in RVC execute by performing a number of discrete computational steps, also referred to as "firings". During each firing, a FUs may:

- Consume tokens from its input ports,
- Produce tokens on its output ports,
- Modify its internal state (if it has any).

A FU may contain memory that is used to store local state. An important guarantee of the RVC computation model is that this state not be shared with other FUs, i.e. FUs communicate with one another exclusively through passing tokens along dataflow connections, and not through shared state. This eliminates race conditions between FUs, and makes RVC specifications executable programs more resilient to the influences of different scheduling policies that can be employed in implementations.

A firing, once initiated, must terminate irrespective of the environment of the FU, i.e. all conditions necessary for its termination (such as the presence of sufficient input tokens) must be ascertained before the firing is begun. Between firings, the FU is in quiescence, i.e. it does not change its state, does not consume tokens, and does not produce tokens.

A simple example of a RVC-CAL FU is shown in the `Add` FU provided below, which has two input ports `t1` and `t2`, and one output port `s`, all of type `int` [5, 26]. The FU contains one *action* that consumes one token on each input ports, and produces one token on the output port. Actions define what happens during the firing of a FU, and also when a firing may occur. In this case, the declaration of input token variables implies that there be one token on both ports `t1` and `t2`.

```
actor Add () int t1, int t2 ⟹ int s :
  action [a], [b] ⟹ [a + b] end
end
```

FUs may include state variables, such as the variable `sum` in the `Sum` FU below. If they do, then actions may modify those state variables in their *bodies*, which are fragments of conventional imperative code between the `do` and `end` keywords:

```
actor Sum () int t,⟹ int s :
  int sum := 0;
  action [a] ⟹ [sum]
  do
      sum := sum + a;
  end
end
```

A FU may have any number of *actions*. The `Merge` FU below has two, each of which copies a token from one of the two inputs to the output port.

```
actor Merge () int Input1, int Input2 ⟹ int Output:
  action Input1: [x] ⟹ [x] end
  action Input2: [x] ⟹ [x] end
end
```

Should a token be present on both input ports, either action may fire. The FU above does not prescribe any policy for choosing between the actions should they both be firable, and consequently its result could be non-deterministic in different implementations of the program. While occasionally useful, non-determinism is a property that FU writers need to be aware of and able to control, which is why many constructs of the RVC-CAL language deal with ways to specify action selection and constraining action firability, and they permit tools to detect potential non-determinism.

One mechanism for picking actions is to use *guards*, conditional expressions associated with an action. The `Select` FU below reads and forwards a token from either port `A` or `B`, depending on the value of the token read from the S input port. That value is tested in the guards of the actions.

```
actor Select () boolean S, int A, int B ⟹ int Output:
  action S: [sel], A: [v] ⟹ [v]
  guard sel end

  action S: [sel], B: [v] ⟹ [v]
  guard not sel end
end
```

In general, guards are arbitrary Boolean expressions that may refer to input tokens and state variables. Another way to control action selection is to order action with respect to their *priority*. In the `BiasedMerge` FU a priority order is established between the two actions labeled `A` and `B`. It ensures that in case both actions can otherwise fire, the one labeled `A` will be given preference over the one labeled `B`.

```
actor BiasedMerge () int Input1, int Input2 ⟹ int
  Output:
  A: action Input1: [x] ⟹ [x] end
  B: action Input2: [x] ⟹ [x] end
  priority A > B; end
end
```

For an in-depth description of the RVC-CAL language, the reader is referred to the standard specification ISO/IEC 23001–4 and to [5] for the full CAL specification. A large selection of FUs building MPEG-2, MPEG-4 SP and AVC decoders is available in the RVC reference software. As it can be understood by these simple examples RVC-CAL is a language that directly provides:

- Encapsulation properties necessary for building a library based specification (composability).
- Expressiveness for specifying data dominated applications such as video coding (i.e. built in operators for naturally describing data streaming/processing systems).

Beside the properties required by the RVC framework, the specification of a video codec in RVC, RVC-CAL FU code plus the network configuration in FNL XML code, results to be more compact on about one order of magnitude in terms of lines of code, than the corresponding reference software. If compared to optimized versions there is still a factor of more than two of advantage without considering that a RVC decoder specification is modular and intrinsically concurrent, thus presenting much richer and useful features versus the classical sequential specifications.

12.1.4.5 The Abstract Decoder Description (ADM)

The Abstract Decoder Model is the conformance point of a RVC decoder specification. The ADM consists of a (sub)set of FUs from the RVC toolbox configured according to the connections between their input and output ports as described by the FNL specification. So the ADM provides a executable behavioral model in the form of a dataflow program that constitutes the unique conformance point of a RVC specification. The program is labeled as abstract because it specify only the minimal elements necessary for a correct execution of a decoding process (the I/O behavior of FUs and the constraints of the RVC computation model). Starting from it is up to the users to derive the implementations using whatever method and implementation language that they prefer starting from such conformance point. particularly attractive are the non normative tools that can synthesize C/C++/Java or HDL directly from the ADM specifications. More details and the SW of these tools can be found in Refs. [6, 11, 12, 24–27] and related references.

12.1.4.6 The RVC Video Tool Library (VTL)

The video tool library is available in ISO/IEC 23002-4 in textual form, supporting the reference RVC-CAL reference software, characterizing the external observable I/O behavior of each FUs in normative form. Therefore, such part of the standard only specifies in a normative way how the values of input data (tokens) are transformed into other values by the algorithm that is encapsulated in the given FUs. However, it must be noticed that such normative behavior is not sufficient to generate a unique executable specification of a codec configuration, such as the one provided by the ADM, because the textual description does not include the *firing rules* of FUs that are necessary to yield a consistent and deterministic system that satisfy the RVC model of computation.

12.1.4.7 The Reference Software Model (RSM)

The reference SW is the RVC-CAL implementation of all FUs constituting the RVC video tool library. Instantiations of these modules plus the specification of a codec configuration builds the executable ADM which is the core of the RVC specification.

12.1.5 Planned Extensions and Perspective

The RVC framework has been recently standardized and the coverage of all existing standard technology is going to be completed in 2011. Next step, after the full standardization of the framework and complete coverage of MPEG video technology, is to be able to timely provide the specification of new standards under development such as HEVC during its definition, thus provide its specification in RVC form together with the final standardization. Other extensions planned are to support more dynamic scenarios in which codec descriptions are associated to content and decoders are instantiated on the fly [28].

References

1. C. Lucarz, M. Mattavelli, J. Thomas-Kerr, and J. Janneck, "Reconfigurable Media Coding: A New Specification Model for Multimedia Coders," in IEEE Workshop on Signal Processing Systems, SiPS 2007, (Shanghai, China), April 17–19, pp. 481–486, 2007.
2. Ihab Amer, Christophe Lucarz, Marco Mattavelli, Ghislain Roquier, Mickael Raulet, Olivier Deforges, Jean-Francois Nezan: "Reconfigurable Video Coding: the Video Coding Standard for Multi-core Platforms", IEEE Signal Processing Magazine, Special issue on Multicore Platforms, Vol 26, No. 6, November 2009, pag 113–123.
3. Marco Mattavelli, Ihab Amer and Mickael Raulet: "The reconfigurable video coding standard", IEEE Signal Processing Magazine, Vol 27, No. 3, May 2010, pag 159–167.
4. Jörn W. Janneck, "Actors and their composition", Formal Aspects of Computing, Vol. 15, No. 4, December 2003, pp. 349–369, Springer.
5. J. Eker and J. Janneck, "CAL Language Report," Tech. Rep. ERL Technical Memo UCB/ERL M03/48, University of California at Berkeley, Dec. 2003.
6. "Open DataFlow Sourceforge Project," http://opendf.sourceforge.net/.
7. Endri Bezati, Marco Mattavelli, Mickael Raulet: "RVC-CAL dataflow implementations of MPEG AVC/H.264 CABAC decoding", Proceedings of the 2010 Conference on Design and Architectures for Signal and Image Processing, DASIP 2010, October 26–28, 2010, Edinburgh, United Kingdom.
8. M. Wipliez and M. Raulet, "Classification and transformation of dynamic dataflow programs," in Design and Architectures for Signal and Image Processing (DASIP), 2010, pp. 303–310.
9. M.Wipliez, G. Roquier, M. Raulet, J.-F. Nezan, and O. D´eforges, "Code generation for the MPEG reconfigurable video coding framework: from CAL actions to C functions," in IEEE International Conference on Multimedia & Expo (ICME), (Hannover, Germany), 2008.
10. Matthieu Wipliez, Ghislain Roquier, and Jean-Franois Nezan, "Software Code Generation for the RVC-CAL Language," Journal of Signal Processing Systems, 2009.
11. J. W. Janneck, I. D. Miller, D. B. Parlour, G. Roquier, M.Wipliez, and M. Raulet, "Synthesizing hardware from dataflow program: an mpeg-4 simple profile decoder case study," in Proceeding of the 2008 IEEE Workshop on Signal Processing Systems (SiPS), November 2008.
12. Jorn W. Janneck, Ian D. Miller, David B. Parlour, Ghislain Roquier, Matthieu Wipliez, and Mickael Raulet, "Synthesizing Hardware from Dataflow Programs," Journal of Signal Processing Systems, 2009.
13. Ab Al-Hadi Ab Rahman, R. Thavot, M. Mattavelli, P. Faure: "Hardware and software synthesis of image filters from CAL dataflow specification", Ph.D. Research in Microelectronics and Electronics (PRIME) 2010 Conference, 18–21 July 2010, Berlin, Germany, ISBN: 978-1-4244-7905-4, Pag. 1–4.

14. G. Roquier, C. Lucarz, M. Mattavelli, M. Wipliez, M. Raulet, J. W. Janneck, I. D. Miller, D. B. Parlour: “An integrated environment for HW/SW co-design based on a CAL specification and HW/SW code generators”, Proceeding of ISCAS 2009, Taipei, May 2009, 978-1-4244-3828-0/09/$25.00 ©2009 IEEE.
15. Richard Thavot, Ab Rahman, Ab Al Hadi Bin, Romuald Mosqueron, and Marco Mattavelli, “Automatic multi-connectivity interface generation for system designs based on a dataflow description,” in 6th Conference on Ph.D. Research in Microelectronics & Electronics, 2010.
16. Christophe Lucarz, Ghislain Roquier, Marco Mattavelli: “High level design space exploration of RVC codec specifications for multi-core heterogeneous plaforms”, Proceedings of the 2010 Conference on Design and Architectures for Signal and Image Processing, DASIP 2010, October 26–28, 2010, Edinburgh, United Kingdom.
17. Shuvra S. Bhattacharyya, Gordon Brebner, Johan Eker, Jorn W. Janneck, Marco Mattavelli, Carl von Platen, and Mickael Raulet: “OpenDF – A Dataflow Toolset for Reconfigurable Hardware and Multicore Systems”, ACM SIGARCH Computer Architecture News, Special Issue: MCC08 – Multicore Computing 2008, Volume 36, Number 5, pp 29–35, December 2008.
18. C. Lucarz, M. Mattavelli, M. Wipliez, G. Roquier, M. Raulet, J. Janneck, I. Miller, and D.Parlour, “Dataflow/actor-oriented language for the design of complex signal processing systems,” in Proceedings of the 2008 Conference on Design and Architectures for Signal and Image processing (DASIP), November 2008.
19. M. Mattavelli, M. Raulet, J. Janneck: “MPEG Reconfigurable Video Coding” in “Handbook of Signal processing Systems”, Pag. 43–68, S. S. Bhattacharyya, Ed F. Deprettere, R. Leupers and J. Takala Editors, Springer 2010 http://www.springer.com/engineering/signals/book/978-1-4419-6344-4.
20. J. Thomas-Kerr and I. Burnett and C. Ritz and S. Devillers and D. De Schijver and R. Van de Walle, “Is That a Fish in Your Ear? A Universal Metalanguage for Multimedia,” IEEE Multimedia, vol. 14(2), pp. 72–77, 2007.
21. J. Thomas-Kerr, J. Janneck, M. Mattavelli, I. Burnett, and C. Ritz, “Reconfigurable Media Coding: Self-Describing Multimedia Bistreams,” in IEEE Workshop on Signal processing Systems SiPS 2007, (Shanghai, China), April 17–19, 2007.
22. Mickaël Raulet, Jonathan Piat, Christophe Lucarz, Marco Mattavelli: “Validation of Bitstream Syntax and Synthesis of Parsers in the MPEG Reconfigurable Video Coding Framework”, 2008 IEEE Workshop on Signal Processing Systems October 8–10, 2008, Washington, D.C. Metro Area, U.S.A, IEEE Catalog No.: CFP08SIG-CDR ISBN: 978-1-4244-2924-0 ISSN: 1520–6130.
23. Christophe Lucarz, Jonathan Piat, Marco Mattavelli: “Automatic synthesis of parsers and validation of bitstreams within the MPEG Reconfigurable Video Coding Framework”, Journal of Signal Processing Systems, 2009, DOI – 10.1007/s11265-009-0395-7 Link - http://www.springerlink.com/content/ gg261p657x483m91.
24. “Open RVC-CAL Compiler,” http://orcc.sourceforge.net/.
25. “OpenForge,” https://sourceforge.net/projects/openforge/.
26. Jorn Janneck, Marco Mattavelli, Mickael Raulet and Matthieu Wipliez: Reconfigurable Video Coding - a Stream Programming Approach to the specification of New Video Coding Systems”: Proceedings of the 2010 ACM Conference on Multimedia Systems (MMSYS 2010), Feb 22–23 2010, Phoenix, AZ, USA, pag 223–234, 2010, ISBN: 978-1-60558-914-5.
27. Shuvra S. Bhattacharyya, Johan Eker, Jorn Janneck, Christophe Lucarz, Marco Mattavelli, and Mickael Raulet, “Overview of the MPEG Reconfigurable Video Coding Framework,” Journal of Signal Processing Systems, 2009.
28. Jérôme Gorin, Matthieu Wipliez, Françoise Prêteux and Mickaël Raulet, “LLVM-based and scalable MPEG-RVC decoder”, Journal of Real-Time Image Processing, Vol. 6, No. 1, pag: 59–70, Springer Berlin / Heidelberg, 2010, url: http://dx.doi.org/10.1007/s11554-010-0169-2.

Chapter 13
MPEG Video/Audio Quality Evaluation

Vittorio Baroncini and Schuyler Quackenbush

13.1 Introduction

Since its very beginning MPEG has been an evolutionary melting pot, where experts push beyond the goal of creation of international standards for new audio and video compression technologies.

This strong push towards innovation had since the beginning characterized the work in MPEG, where a tight schedule is "mandatory" for any action undergoing. The schedule has to be tight also when a new technology has to be considered, evaluated and demonstrated at the end of its standardization process. An important part of this schedule is certainly given by the testing activities; testing in MPEG means formal and informal subjective assessment, because the eyes of experts are an essential tool to assess the improvements and the eyes of naïve viewing subjects are the "real" tool to provide a formal proof of these improvements.

So tight schedule and need to proof the technologies are key parts of the MPEG story; but what happens when a technology needs new testing tools to be proven? MPEG moves ahead also in the area of formal (and informal) subjective assessment! Because even if objective measures (e.g. PSNR) provide a good indication of the improvements, the final judgement can be given only by the human eyes.

The MPEG schedule is typically done of a "Call for Evidence" (where the new ideas are pre-screened to see if the way to a new standard is worth), a "Call for Proposals" (the "real competition" in MPEG, open to external parties also) and the "Verification tests" (the "showcase" of the new technologies). All these steps have in common the formal subjective assessment of the video or audio, always done in a professional and following the existing standards; but it may happen that MPEG schedule

V. Baroncini (✉)
Fondazione Ugo Bordoni, Viale del Policlinico 14700161, Rome, Italy
e-mail: vittorio@fub.it

L. Chiariglione (ed.), *The MPEG Representation of Digital Media*,

DOI 10.1007/978-1-4419-6184-6_13,

asks for quick answers and flexibility; therefore when the schedule does not allow proceeding in a "standard" way (i.e. following the existing Recommendations), MPEG reinvents (thanks to the many and authoritative experts operating in the Test sub-Group) the rules of formal and informal subjective assessment. It also happened that the Recommendations in force were improved thanks to the MPEG testing activities and innovations.

This chapter tries to describe the evolution of formal (and informal) subjective assessment in MPEG, and how it influenced the "best practice" in the area of subjective assessment. An example of this influence is the use in video testing of "expert viewing" trials; the MPEG "collaborative phase" implies the creation of several "core experiments" in which the participants submit their proposals and a comparison is made to select the better proposals; this process is mainly made of "objective" evaluation of the results (e.g. PSNR for video). But it is recognised in video testing that "visual assessment" could be relevant; to do this a limited number of experts are used to provide their opinion according to a rather stable "expert viewing" protocol "informally" adopted in MPEG, also if with some "variations" dictated by case by case conditions. In audio testing, subjective listening tests are the norm, as common objective measures (e.g. SNR) are well known to have at best modest correlation to subjective quality. This "best practice" has not been adopted outside of MPEG, even if several attempts were done to insert it the existing standard literature (e.g. ITU-R Recommendations), and even if it is regularly used in several research contexts.

MPEG's influence on formal subjective testing was strong also for the statistical analysis of the subjective data. The use of the Confidence Interval (CI) and the analysis of "disaggregated data" for each test sequences was introduced in MPEG and then became a consolidated "best practice" also outside MPEG. Furthermore the new standard methods, currently part of the relevant literature of other International Bodies came quite exclusively from testing activities done in MPEG.

13.2 MPEG Subjective Audio Quality Assessment

As in MPEG Video standardization, MPEG Audio uses a similar process in developing new standards. The critical step prior to embarking on a standardization effort is to determine if new technology has performance that is significantly better than existing MPEG audio standards. This is done via issuing a Call for Evidence (CfE). Such a document indicates the functionality and compression performance required for a possible new standard and also clearly describes the process and criteria to be used in evaluating responses to the Call. Perhaps the most important component in the evaluation process is a formal subjective evaluation of the performance (e.g. subjective quality at a given transmission bitrate) of the submitted technologies.

Such formal subjective tests use well-recognized and widely-adopted international standards for subjective assessment, such as ITU-R Recommendation BS-1,116 (triple stimulus hidden reference) [1] for technology with very small audio impairments, or ITU-T Recommendation BS.1534-1 (MUHRA) [2]. Each of these test methodologies specify that listening be done with highest-quality audio equipment in acoustically controlled environments and with expert listeners. The last point cannot be over-stated: expert listeners are trained, either via a lengthy training process prior to the subjective test or via extensive professional experience, such that they are highly adept at discerning the relevant audio impairments in a testing scenario. This leads to more stable test results and requires that fewer subjects participate in the test.

If the results of the CfE subjective tests and other expert evaluations (e.g. of technology functionality) are such that all MPEG experts agree that there exists new technology that meets the requirements for a possible new standard, then the next step is to issue a Call for Proposals. This is typically a public call and contains the same elements as a CfE: a clearly described set of functionality and performance requirements, a detailed timeline for the evaluation process and a detailed description of the evaluation process, including the subjective assessment of the proposed technologies. Based on the evidence associated with the Call, a "best" technology is selected as the "Reference Model" (RM). This concludes the competitive phase of the standardization process and the RM is then the basis for the subsequent collaborative phase of the standardisation process.

The collaborative phase is defined by the "Core Experiment" (CE) process. Given the RM, a CE proponent proposes technology that replaces an existing tool or adds a new tool to the RM. In order to assure that the performance of the RM only increases during the collaborative phase, each CE must demonstrate significantly improved performance of the new RM configuration ("RM+CE") relative to the original configuration ("RM"). Such performance improvement is typically demonstrated via subjective listening tests, for example showing that RM+CE achieves an average subjective score that is higher than that of RM at the 95% level of significance. Note that some CEs may offer improvements that are not quantified by changes in subjective quality, e.g. lower complexity while achieving identical decoded output waveforms. In such cases listening tests are not a component in evaluating the CE technology.

The collaborative phase is a process that is unique to MPEG and the continuous check on the performance of the incremental changes to the RM is a process unique to MPEG Audio. It is widely agreed by MPEG Audio experts that the performance of audio compression technology can only be reliably assessed via subjective means. There is no reliable objective assessment means, particularly as MPEG standards push to regions of ever higher compression which necessitate parametric coding methods rather than waveform-preserving coding methods.

Just prior to the final stage of standardization, MPEG Audio conducts a formal Verification Test of the standardized technology. The purpose of such a test is to demonstrate to potential customers the performance of the standard and also to confirm that the standard fully satisfies the requirements set forth in the original CfP. A rigorous subjective assessment process, using expert listeners in acoustically controlled environments is essential to having the Verification Test Report be a document that conveys a reliable message and withstands the most rigorous level of scrutiny. Considering the effort put into developing a typical MPEG standard, such effort for a Verification Test is well spent!

13.3 MPEG Subjective Video Quality Assessment: A Continuous Evolution

The beginning of this evolution occurred in the first MPEG-4 Competition Test, when professional TV apparatuses were used to assess the quality of multimedia video; a D1 video tape recorders, video signal distributors and "grade 1" TV studio monitors allowed a lossless reproduction of the video clips; this became the "reference visual evaluation set-up" adopted by MPEG for a long time.

Also from the point of view of the test methodology and statistical analysis of the results those tests represented a milestone. In fact a new test method trying to evaluate the effect of transmission errors on visual quality was designed and the concept of "statistically significantly better" (SSB) was introduced in the evaluation of the results. This improvement was achieved considering the "confidence interval" (CI) of the raw subjective data.

This experience is described in a previous book on MPEG-4 [3] but is important to be remembered as it represented the beginning of a continuous process of innovation in the evaluation of visual quality still in progress in MPEG. Here below the evolution of the formal subjective video quality assessment is described, taking as a reference the changes in the methods and in the laboratory set-up adopted by MPEG to conduct the more important Competition and Verification tests performed during the last years.

13.4 Digital Cinema Competition Test

The Call for Proposal for a compression technology dedicated to Digital Cinema applications was a challenging effort from the logistic point of view and it represented an interesting evolution in the subjective evaluation practice.

At the time of the MPEG D-Cinema test the technology of projection of digital contents in a Cinema Theatre was at an early stage. The laboratory set-up was therefore reinvented from the scratch using a real Cinema Theatre: the Entertainment Technology Centre of the University of South California (Hollywood).

The ETC is located in the historical cinema theatre build by the Warner Brothers, which still now represents one of the best locations for a cinematic representation, due to its elliptic shape that allows a perfect vision of the screen from any seat of the theatre.

Two D-Cinema video servers (AVICA 4:4:4 and DVS 4:2:2 HD video servers) were used with a Christie TI DLP projector (1,920 × 1,080 resolution, with Cinema technology) to illuminate a 24 × 13 ft non perforated high reflective flat screen. The viewing subjects were seated at the 6th and 12th rows of the Theatre, representing a viewing distance of 1,5 and 3 times the screen height respectively.

Two new test methods were introduced for this subjective evaluation: a sequential evaluation (the Double Stimulus Perceived Difference Scale – DSPDS) and a side by side evaluation (Double Stimulus Split-Screen Perceived Difference Scale – DS3PDS) method. The DSPDS method was a derivation of the DSCQS method; the timing of the presentation of the video clips was the same of the DSCQS where the original and coded samples are shown two times without any indication of which of the two is the original. Only one vote was required to the subjects to indicate if they can perceive any difference between the two video clips.

The DS3PDS method was designed splitting the screen in two equal parts to show at the same time the half of coded and half of the original scenes. This method was used when the level of impairments was very low (higher bit rates), to allow a better evaluation of the differences.

Both methods used a 5 levels "difference perception scale" in which the classic five levels ITU quality adjectives (excellent, good, fair, poor, bad) were replaced by five adjectives retained to be of higher semantic meaning: "No visible difference" "Slightly visible differences" "Some easily visible differences" "Many easily visible differences" and "Overall visible differences". This choice was suggested by some negative comments on the "real" meaning of the ITU quality adjectives. The used adjectives are similar to the ITU "Impairment Scale" reported in Ref. [4], but modified to be more adherent to the test context.

A minor but useful logistic solution was the use of "light pens" to allow to the viewing subjects to fill out the scoring sheets in a complete dark ambient (like a cinema theatre is). The "light pen" was a nice gadget, represented by a ball point pen with a LED light on the writing pin side, which illuminated just the paper sheet when pointed on it to write.

An important aspect of the laboratory set-up was the measurement of the illumination of the screen and the set-up of the projector; both required one full day to be completed. Details of these procedures and of the whole test history can be found in Ref. [5].

13.5 AVC Verification Test

Let's now consider the test performed for the MPEG-4 Part 10 (AVC) technology. The "Call for Evidence" and the "Call for Proposal" for the new AVC standard were done using a professional studio TV set-up. It was for the "Verification Tests" of AVC (December 2003) that the professional TV laboratory set-up, adopted in 2001, was partially abandoned. The reason for this change was in the wide range video formats considered in the Verification Test; the target application of the new compression technology was in fact covering a very wide range of image dimensions (from QCIF to full HDTV), of bit rates (from 10 Kbps up to 20 Mbps) and frame rates (from 10 to 30 frames per second).

The above situation was very demanding in the design of the test and in the definition of the laboratory set-up. For the first time in the formal subjective assessment of video two of the four subjective test experiments (the MD Baseline and the MD Main tests) were conducted using a PC connected to a 19″ LCD monitors; the

PC used was the most powerful available at that date[1] and the monitor resolution was set to 1,024 pixels and 768 lines.

The above decisions were taken considering that the new compression technology had been developed having in mind a Multimedia environment, i.e. a user sitting at a desk and using the computer; it was therefore almost natural to run the subjective quality evaluation sitting the viewers in front of a PC. The improvement in the laboratory set-up was allowed by the low computational power required to play video clips of small size, such as those used in the MD Baseline and MD Main tests (QCIF and CIF[2]); furthermore the active part of the LCD screen (i.e. the part of the screen actually representing the video) was so small that only one subject at a time could be seat in front of the monitor at a distance of 4H from the screen (where H is the height of the active part of the screen). This led to run the test connecting four LCD displays to a PC to minimise the time required to complete the assessment trial.

Also for the HD Main Tests the laboratory set-up was improved using a Christie DLP S3000 projector (720p and the 1,080 25p video material) and a Panasonic 30″ CRT broadcast monitor (1,080 60i). The viewing distance was set at 2H for the DLP projector and at 3H for the CRT. The peak brightness (full screen white) of the projection was 12 fL. Ambient illumination was made of scattered light from the projector. For the CRT, back ground was illuminated with D-65 fluorescents, 10% less bright of the screen. Viewers were provided with LED white light pens to fill out the scoring sheets. A full description of the MPEG-4 Part 10 verification test can be found in Ref. [6].

13.6 Scalable Video Coding

13.6.1 The "Call for Evidence" Test

The "evidence" of the SVC technology was done during the meeting with the participation of MPEG experts. The test was run in a conference room, properly obscured from external light sources and far from any external audible pollution; a 42″ 16:9 Plasma monitor (Sony PFM-42B1E) and an high performance PC were used to display the video test material; five experts were seated at 4H and five were seated at 5H from the screen in a low illuminated room. The experts carried out the whole test, having 5 min time of rest every hour.

The SVC proposals were compared to each other and the best one was compared to an AVC anchor. The comparison was made in a side by side (split screen) mode,

[1] Pentium IV 1,2 GHz – 1G RAM – 512M dual head Video Board – 1T HDD.

[2] The CIF (Common Interchange Format) video format is made of moving images with 352 columns and 288 lines; QCIF (Quarter Common Interchange Format) video format is made of moving images with 176 columns and 144 lines.

and the experts knew where AVC and SVC were located; each test case was shown in a loop mode three times; at the end of the presentation the question "is left or right better?" was asked, followed by a short discussion on how much SVC was "better" (or worse) to AVC.

Comments were collected on the site set-up and the answers showed that the monitor position and in general the accommodation were retained good; also the voting procedure was generally judged satisfactory. The report of the call for evidence testing activity can be found in Ref. [7].

13.6.2 The "Competition" Test

The Competition test of the SVC technology involved the evaluation of video material that, for the lower resolutions, was in absence of an unimpaired reference; this situation demanded for the usage of a absolute category rating test method, i.e. the Single Stimulus Multi Media (SSMM) test method. The SSMM is derived from the SS method [4], showing two times (in the same session) the video clips, to minimize the "contextual" effect. SSMM uses an eleven grade numerical quality scale, as suggested by Annex B ("Additional evaluative scales") to Ref. [8].

An important improvement was done in the laboratory set-up, were for the first time a PC, multimedia displays and "Home Theatre" DLP projectors were used. The availability of RAID[3] HDD controllers improved dramatically the performances of the Hard Disk Drives when reading and playing video, allowing a smooth playback.

It is useful to remember that when the viewing distance from a Professional TV monitor is equal of inferior to 3 times the screen height, the interlaced structure of the monitor does not allow a correct vision of progressively scanned video clips. For this reason 19″ computer LCD displays were used for the high resolutions test (quad-CIF), and a DLP projector was used for the low (QCIF) and medium (CIF) resolution tests.

The use of a DLP projector provided also the advantage to show the CIF and QCIF video clips in a relatively big size (20″ and 10″ respectively); this allowed to observe the video at a 4H distance without seating the subjects too close to the screen and to a rather constant viewing distance. We should consider that on a 19″ LCD display a CIF image is 10 cm high, leading to a 4H viewing distance of 40 cm; at that distance any small movement of the viewers' head drastically modifies the viewing distance, and therefore the level of visibility of the image impairments. The complete description of the SVC competition test can be found in Ref. [9].

[3] Redundant array of independent disks.

13.6.3 The Verification Test

The verification of the Scalable Video Coding technology was based on the experience gained during the Competition Test; furthermore a new test method, the Double Stimulus Unknown Reference (DSUR), was introduced to evaluate the higher resolution video. The laboratory set-up was the same already well verified during the Competition Tests. Also in this case a DLP projector was used for the lower resolution video clips, while CRT and LCD monitors were used for the higher resolution video clips. The more interesting part of this test experiment was related to the use of the new DSUR test method for the evaluation of the high resolution video clips (720p and 1080i), where an unimpaired reference was available. This method is based on the same timing of the DSCQS test method; the novelty is represented by the kind of task required to the subjects and by a different the voting scale. The DSCQS method requires for the subjects to score separately the two video clips shown two times during a basic test cell; the viewers do not know the order of presentation of the original and of the coded video clips; furthermore the order of presentation itself is randomly changed at each test cell.

The DSUR adopts the same presentation of the original and coded video clips, but the viewing subjects are asked to perform a different task that includes a small cognitive effort. They are asked to use the first pair of presentations to decide which of the two video clips is the original, and the second pair of presentations to score the coded clips against the one they have decided to be the original. The cognitive effort helps to maintain a good level of attention during the whole test sessions. More details about the SVC verification test and on the DSUR test method can be found in Ref. [10].

13.7 High Efficiency Video Coding

13.7.1 The "Call for Evidence" Test

The test plan of the "Call for Evidence" (CfE) of a High Efficiency Video Coding (HEVC) foresaw the conduction of a formal subjective assessment test to be carried out during the 90th MPEG London meeting, with the participation of a selected number of MPEG experts. The test plan included the adoption of the DSCQS test method, to compare the Anchor (i.e. the previous technology, in this case AVC) with the responses to the call. The 20 submissions received in response to the CfE made not feasible to perform a complete DSCQS subjective test, during the 5 days of the MPEG meeting (a raw computation of the required time indicated more than 10 working days of continuous testing activity).

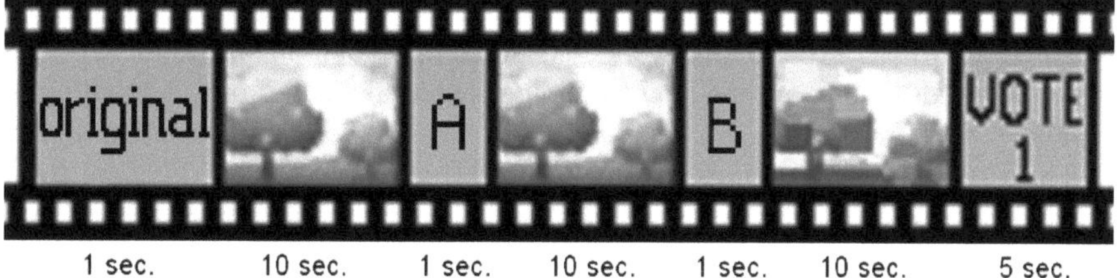

Fig. 13.1 Expert viewing basic test cell

A new "expert viewing" protocol was therefore designed by the test manager, allowing the evaluation of the proposals into two days. The protocol adopted was relatively simple and it is based on the following points:

- Three experts participated to the assessment;
- Each test case was compared to the original and to the Anchor;
- The experts evaluated the submission using an 11 levels quality scale. This led to the design of a BTC[4] as depicted in Fig. 13.1.

To avoid any bias in the evaluation of the experts, the order of presentation of the Anchor (A or B) and of the submission (B or A respectively) was randomised.

The laboratory set-up was done seating the three subjects at 1,5H form a 30″ high quality LCD display by EIZO (CG301W). Ambient light was low and the room was far from any visible or audible external pollution. The player was the same already used for previous subjective assessment trials in MPEG, but the dimension of the video files required the use of a extremely powerful PC based on a Pentium IV CPU, four high speed SATA disks (RAID 0) and two 1G RAM video boards in CrossFire configuration. The connection between the video boards and the display was done using a dual link DVI video cable. The results were obtained averaging the scores of the three experts and no statistical processing was made. A comparison between the subjective results and the PSRN values, showed a very good performance of the expert viewing procedure. The complete description of the HEVC CfE test can be found in Ref. [11].

13.7.2 The "Competition" Test

The Competition test of the HEVC technology was the biggest subjective assessment effort ever done in the history of the formal video quality evaluation, not only in MPEG but in any other international standardization body. In fact in response to the Call for Proposal MPEG received a total of 27 submissions, each of them fully

[4] BTC – Basic Test Cell; the core element of a test session in which a coding condition is evaluated.

covering the requested 116 coding conditions. The test included also two Anchors prepared by MPEG experts, that were considered as regular answer to the CfP and that were used to compare the new technology with the exiting one. According to the above, the Test Coordinator designed a test to assess 29 complete submissions.

The 116 coding conditions foreseen by final version of the CfP, came from 4 video classes (Class B, Class C Class D and Class E), 4 video clips (5 for Class B), 4 coding bit rates and two coding modes ("random access" and "low delay"); his led to a total of $29 \times 116 = 3{,}364$ test cases (each made of a 10 s long video clip).

Each Proponent delivered to the Test Coordinator an HDD with 80 Gbytes of video data; several HDD suffered of delivery problems (broken, late delivery, wrong content) and a long and patient negotiation was undertaken to get the complete and correct submission from all the 27 Proponents; at the end more than 40 HDD drives were sent to the Test Coordinator to achieve a complete reception of the 27 submissions. The Test Coordinator (and his team) did a careful visual check of all the video clips received by the Proponents, that, considering the also multiple submissions (due to delivery problems, file creation errors and HW problems) lead to more than 4,000 video files to be verified, before being distributed to the three test laboratories. The delivery of the test material to the test laboratories was done by the Test Coordinator, bringing in person two copies of the full set of the received submissions, on a package of 4 TByte HDDs.

For three of the four classes (Class B, Class C and Class D) the CfP required to assess together video clips with a different frame rate. This constrain did not allow to use commercially available video server, but required the use of a dedicated player running on a extremely powerful PC. The huge size of the video files required the availability of an extremely high speed storage system to perform a smooth playback of the video clips; this led to the usage of a new storage technology: the SSD[5] devices. Furthermore the play-out of HD video files @ 60 fps required the use of the most reliable and high speed video board available on the market in SLI or X-Fire configuration coupled to powerful motherboard. The three test laboratories involved in the test were equipped with similar HD and SW system; all of then were configured with a back-up playback system.

The total number of test conditions was extremely high (3,364) and had a strong impact on the design of the test. The design of a formal subjective assessment of 3,364 test cases has been never done before in any laboratory across the world. This was known since the very beginning to be a very challenging effort, which had to be undertaken being aware of the many risks of failure. Furthermore the 10 weeks schedule was very tight and left very short time to any recovery actions. For this reason the Test Coordinator asked to the MPEG community a "priority list" in the case it was not possible to complete the whole process during the 12 weeks available between the 91st and the 92nd MPEG meeting. The priority was given ranking

[5] SSD – Solid state Storage Device; the new disk made of flash memory with an extremely low data read time; four SSD drives in Raid 0 (stripe) configuration provide a read speed four times higher than the equivalent configuration using conventional HDD.

Class B and Class C as the most important and giving to Class E and D the lowest priority. Nevertheless all the test were completed and the results were available at the HEVC AhG meeting, scheduled for the Sunday before the 92nd MPEG meeting.

The design of the test was the most challenging part of the whole experiment. The most important requirement in the design was the full randomization of the test cases across all the test sessions (and inside each test session). Another mandatory constrain was the presence of the anchors in each test session. Furthermore the video clips form each proponent had to be homogeneously distributed over the whole test sessions. The design was done individually for each video Class and led to 136 test sessions; each test session was properly randomised and completed with a "stabilization phase", made of a set of three test cases done selecting the ones showing high, medium and low quality; this procedures is useful to provide (as recommended in Ref. [4]) to the viewing subjects an immediate indication of the range of quality they are going to evaluate for that test session.

The test methods used in this experiment were the DSCQS and the DSIS (with a 11 levels quality scale). The DSCQS method was adopted only for the Class B, and in particular only for the lower compression rate points; in those cases the impairment introduced by the encoding process was retained to be low and therefore requesting the adoption of a more accurate visual estimation. For all the other cases and classes the DSIS test method (Variant I) was used allowing a huge saving of time and resources.

The test required the participation of more than 800 viewing subjects, screened for visual acuity and colour vision and carefully selected among University students. All the subjects were properly trained; particular care was given to the meaning of the 11 levels of the quality scale used to rate the video in the DSIS method. More in detail it was explained that the vote 10 had to be used when that they saw no difference between original and processed images; 9 and 8 had to used when very small differences were seen; 7and 6 when few impairments were visible but paying a lot of attention; 5 and 4 when some impairments were easily visible; 3and 2 when impairments were easily visible all over the image; 1 and 0 when many impairments were visible without any difficulty.

The data processing of the results was done computing the MOS and the CI values. It must be pointed out that the graphical representation of the 3,364 test cases represented one of the more difficult aspects of the data processing phase; in fact it was necessary to provide a rather simple representation of the ranking of the Proponents for each Test sequence, for each Class and for each coding condition.

A solution was found creating a graph (see Fig. 13.2) for each test sequence, for each Class and for each Mode, in which all the scores of all the Proponents (including the Anchors) were represented ordering the points of the graph for ascending quality and for one compression bit rate.

The presence of labels on the X axis allowed to compare at glance the Proponents and to rank them; the CI value was added centring it on the MOS value, to provide an indication of the cases where "statistical difference" was found. There was a request of confidentiality of the results of the test from the Proponents; this was solved assigning a secret code to each proponent, allowing them to know their performances maintaining the due privacy. The results of the HEVC test were presented

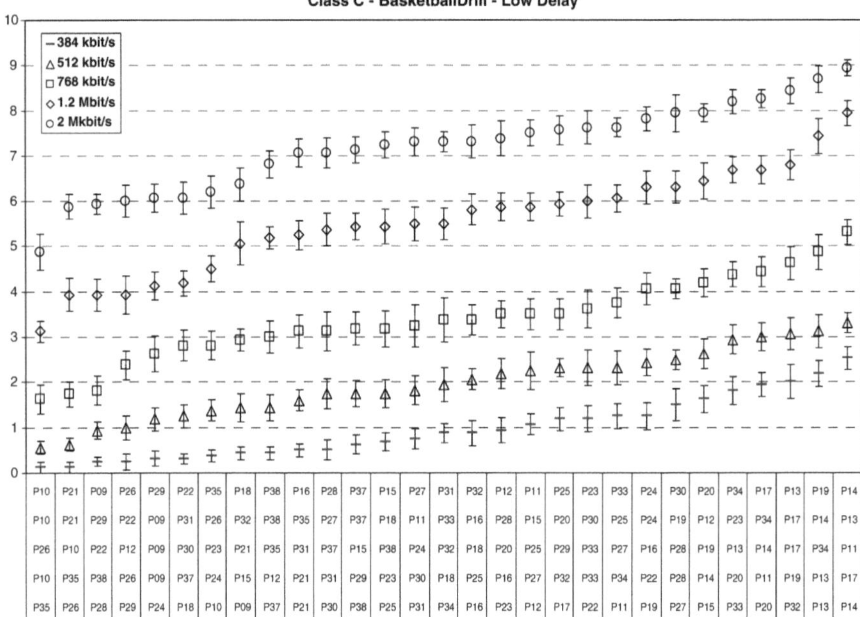

Fig. 13.2 Example of graphical representation of the test results (Class C, test sequence "Basketball Drill" and coding mode "Low Delay")

at the AhG meeting before the 92nd MPEG meeting and were judged positively by the experts participating to the meeting. More details about the HEVC test can be found in Ref. [12].

13.8 Conclusions

This chapter describes in the evolution of the subjective assessment "best practice" in MPEG, taking into account the experience gained in the exploitation of the Call for Evidence, Call for Proposals, Core Experiments and Verification testing efforts during the last decade.

This evolution was made possible thanks to the participation of many video and audio MPEG experts. The most important aspect of this evolution is given by the complete freedom that the experts had in the definition of new test procedures and laboratory set-up. This made MPEG a place where new ideas were quickly and effectively developed to obtain the necessary results. Such a freedom was always used together with a solid knowledge and strong background in the area of audio and video quality assessment. The target has always been the achievement of results in the shortest time, optimising the resources and being supported by the highest professionalism and technical expertise. The new rules have always been designed and applied adapting the "classic rules", formally defined in the existing literature, to the needs and the requirements that MPEG had to face time by time.

References

1. ITU-R Recommendation BS. 1116-1, "Methods for the subjective assessment of small impairments in audio systems including multichannel sound systems," Geneva, Switzerland, 1997 (available at http://ecs.itu.ch)
2. ITU-R Recommendation BS.1534-1, "Method for the subjective assessment of intermediate quality level of coding systems: MUlti-Stimulus test with Hidden Reference and Anchor (MUSHRA)," Geneva, Switzerland, 1998–2000. (available at http://ecs.itu.ch)
3. Fernando Perreira and Touradj Ebrahimi editors "The MPEG-4 book"IMSC Press Multimedia Series – Prentice Hall PTR – 2002
4. ITU-R Recommendation BT 500-11, "Methodology for the subjective assessment of the quality of television pictures" Geneva, Switzerland, 1998-2000. (available at http://ecs.itu.ch)
5. MPEG Test and Video. *Results of Assessment of Responses to Digital Cinema Call for Proposals* Doc. ISO/MPEG N4455, Pattaya 2001
6. MPEG Test and Video. *Report of The Formal Verification Tests on AVC* Doc. ISO/MPEG N6231, Waikoloa 2003
7. MPEG Video. *Report on Call for Evidence on Scalable Video Coding (SVC) technology* Doc. ISO/MPEG N5701, 2003, Trondheim
8. TU-T Recommendation P.910, "Subjective video quality assessment methods for multimedia applications" Geneva, Switzerland, 1998–2000 (available at http://ecs.itu.ch)
9. MPEG Test and Video. *Subjective test results for the CfP on Scalable Video Coding Technology* Doc. ISO/MPEG W6383, 2004, Munich
10. Tobias Oelbaum *Subjective results for the SVC Verification Test* Doc. ISO/MPEG M15012, 2007, Shenzhen
11. MPEG Video, Test and Requirements. *Results of Call for Evidence on High-Performance Video Coding (HVC)* Doc. ISO/MPEG N10721, 2009, London
12. MPEG Test, Video and JCT-VC. *Report of Subjective Test Results of Responses to the Joint Call for Proposals (CfP) on Video Coding Technology for High Efficiency Video Coding (HEVC)* Doc. ISO/MPEG N11275, 2010, Dresden

Abbreviated List of Acronyms

(Leonardo Chiariglione – CEDEO.net)

Acronyms	Definition
3DG	3D Graphics
3DV	3D Video
AAC	Advanced Audio Coding
ADM	Abstract Decoder Model
AFX	Animation Framework eXtension
AHG	Ad Hoc Group
AMD	Amendment
API	Application Programming Interface
ASO	Arbitrary Slice Ordering
AU	Access Unit
AVC	Advanced Video Coding
B	Bi-predictive
BC	Backwards Compatible
BD	Blu-ray Disc
BIFS	Binary Format for MPEG-4 Scenes
BP	Baseline Profile
BSDL	Bitstream Syntax Description Language
CA	Conditional Access
CABAC	Context-Adaptive Binary Arithmetic Coding
CAVLC	Context-Adaptive VLC
CBP	Constrained Baseline Profile *or* Coded Block Pattern
CCITT	Consultative Committee for International Telephony and Telegraphy
CD	Committee Draft *or* Compact Disc
CE	Core Experiment
CELP	Code Excited Linear Predictive
CfE	Call for Evidence

L. Chiariglione (ed.), *The MPEG Representation of Digital Media*,

DOI 10.1007/978-1-4419-6184-6,

Acronyms	Definition
CfP	Call for Proposals
CG	Computer Graphics
CGS	Coarse-Granularity Scalability
CIF	Common Intermediate Format
CPB	Coded Picture Buffer
CPS	Constrained Parameter Set
CPU	Central Processing Unit
CVS	Coded Video Sequence
DAM	Draft Amendment
DCT	Discrete Cosine Transform
DERS	Depth Estimation Reference Software
DIS	Draft International Standard
DPB	Decoded Picture Buffer
DPCM	Differential Pulse Code Modulation
DRM	Digital Rights Management
DSM-CC	Digital Storage Media Command and Control
DSP	Digital Signal Processing
DTD	Document Type Definition
DTS	Decoding Time Stamp
DVB	Digital Video Broadcast
DVD	Digital Versatile Disc
ECM	Entitlement Control Messages
EMM	Entitlement Management Messages
EOB	End of Block
EP	Extended Profile
EPG	Electronic Program Guide
ES	Elementary Stream
FCD	Final Committee Draft
FDAM	Final Draft Amendment
FDIS	Final Draft International Standard
FGS	Fine Granularity Scalability
FLC	Fixed-Length Coding
FMO	Flexible Macroblock Ordering
FND	Functional Unit Description
FNL	Functional unit Network Language
FRExt	Fidelity Range Extensions
FTV	Free-viewpoint Television
FU	Functional Unit
GBSC	Group of Blocks Start Code
GN	Group Number
GNU	GNU is Not Unix
GOB	Group of Blocks
GOP	Group of Pictures

Acronyms	Definition
GPL	(GNU) General Public License
GUI	Graphical User Interface
H422P	High 4:2:2 Profile
HD	High Definition
HDL	Hardware Description Language
HDTV	High Definition Television
HE-AAC	High-Efficiency Audio Coding
HEVC	High-Efficiency Video Coding
Hi10P	High 10 Profile
HL	High Level
HoD	Head of Delegation
HP	High Profile
HRD	Hypothetical Reference Decoder
HTTP	Hypertext Transfer Protocol
I	Intra-picture
IAR	Image Aspect Ratio
IDCT	Inverse DCT
IDL	Interface Description Language
IDR	Instantaneous Decoding Refresh
IEC	International Electrotechnical Commission
IETF	Internet Engineering Task Force
IP	Internet Protocol
IPMP	Intellectual Property Management and Protection
IPR	Intellectual Property Rights
IPTV	Internet Protocol Television
IS	International Standard
ISO	International Organization for Standardisation
IT	Information Technology
ITU	International Telecommunication Union
ITU-R	ITU – Radiocommunication Sector
ITU-T	ITU – Telecommunication Sector
JCT	Joint (ITU-T/ISO/IEC) Collaborative Team
JPEG	Joint (ITU-T/ISO/IEC) Photographic Coding Experts Group
JTC 1	Joint (ISO/IEC) Technical Committee 1
JVT	Joint (ITU-T/ISO/IEC) Video Team
LGPL	(GNU) Lesser General Public License
LOD	Level of Detail
LPC	Linear Predictive Coding
LSF	Low Sampling Frequencies
LTE	Long-Term Evolution
M3W	Multimedia Middleware
MB	Macroblock
MBA	Macroblock Address

Acronyms	Definition
MBAFF	Macroblock-Adaptive Frame-Field
MC	Motion Compensation
MDCT	Modified Discrete Cosine Transform
MGS	Medium-Granularity Scalability
ML	Main Level
MMCO	Memory Management Control Operation
MP	Main Profile
MP3	MPEG-1 Audio Layer III
MPEG	Moving Picture Experts Group
MPTS	Multi-program Tranport Stream
MV	Motion Vector
MVC	Multi-view Video Coding
MVD	Motion Vector Difference
MXM	MPEG eXtensible Middleware
NAL	Network Abstraction Layer
NB	National Body
NBC	Non Backwards Compatible
NP	New Project
NTSC	(United States) National Television Systems Committee
OS	Operating System
OSS	Open Source Software
P	Predictive
P2P	Peer-to-Peer
PAFF	Picture-Adaptive Frame-Field
PAL	Phase Alternating Line
PAR	Pel Aspect Ratio
PAT	Program Association Table
PCM	Pulse Code Modulation
PDAM	Proposed Draft Amendment
PES	Packetized Elementary Streams
PHP	Progressive High Profile
PID	Program ID
PMT	Program Map Table
POTS	Plain Old Telephone System
PP	Professional Profiles
PPS	Picture Parameter Set
PS	Program Stream
PSC	Picture Start Code
PSI	Program Specific Information
PSNR	Peak Signal-to-Noise Ratio
PTS	Presentation Time Stamps
QCIF	Quarter-CIF
QoS	Quality of Service

Acronyms	Definition
QP	Quantization Parameter
QVGA	Quarter VGA (320 × 240)
RA	Registration Authority
RAM	Random-Access Memory
RAND	Reasonable And Non Discriminatory
RGB	Red Green and Blue
RM	Reference Model
ROM	Read-Only Memory
RPLR	Reference Picture List Reordering
RTP	Real-Time Protocol
RVC	Reconfigurable Video Coding
RVLC	Reversible VLC
SAOC	Spatial Audio Object Coding
SBC	Subband Coding
SBR	Spectral Band Replication
SCR	Systems Clock Reference
SDM	System Decoder Model
SDTV	Standard-Definition Television
SEI	Supplemental Enhancement Information
SI	Switching Intra-picture
SIF	Standard Image Format
SMR	Signal-to-Mask Ratio
SNHC	Synthetic and Natural Hybrid Coding
SNR	Signal-to-Noise Ratio
SP	Switching Predictive
SPS	Sequence Parameter Set
SPTS	Single Program Transport Stream
STC	Systems Time Clock
STD	Systems Target Decoder
SVC	Scalable Video Coding
SVG	Simple Vector Graphics
TMUC	Test Model Under Consideration
TR	Temporal Reference
TS	Transport Stream
TV	Television
UHDTV	Ultra-High Definition Television
URQ	Uniform Reconstruction Quantizer
USAC	Unified Speech and Audio Coding
VBV	Video Buffer Verifier
VCEG	Video Coding Experts Group
VCL	Video Coding Layer
VGA	Video Graphics Array (640 × 480)
VLC	Variable-Length Coding

Acronyms	Definition
VO	Video Object
VOP	Video Object Plane
VRML	Virtual Reality Modeling Language
VSRS	View Synthesis Reference Software
VUI	Video Usability Information
W3C	World Wide Web Consortium
WD	Working Draft
WG	Working Group
XML	eXtensible Markup Language
YCbCr	Luminance (Y), Chrominance-Blue, and Chrominance-Red

MIX
Papier aus verantwortungsvollen Quellen
Paper from responsible sources
FSC® C105338

If you have any concerns about our products,
you can contact us on
ProductSafety@springernature.com

In case Publisher is established outside the EU,
the EU authorized representative is:
Springer Nature Customer Service Center GmbH
Europaplatz 3, 69115 Heidelberg, Germany

Printed by Libri Plureos GmbH
in Hamburg, Germany